FUNDAMENTALS OF CANDU REACTOR PHYSICS

Wei Shen
Benjamin Rouben

Library of Congress Cataloging-in-Publication Data

Names: Shen, Wei (Reactor physicist), author. | Rouben, Benjamin (Reactor physicist) 1944- author.

Title: Fundamentals of CANDU reactor physics / Wei Shen, Benjamin Rouben.

Description: New York : ASME Press, 2021. | Includes bibliographical references. | Summary: "Most textbooks on reactor physics focus either on application to Light Water Reactors (LWRs) or on the theory and mathematics, with a significant number of equations and computational schemes. Many were written more than 20, or even more than 60, years ago, and therefore they do not reflect the evolution of reactor concepts and engineering requirements since then. Those categories of books either are difficult to follow for non-physicists working in the nuclear indus-try, or else are of little value for those who are interested in special features of CANDU reactor physics. With more than 75 combined experience in the area of CANDU and LWR reactor physics and safety, the authors provide a practical book on reactor physics for CANDU reactors, particularly for the safe operation of aged CANDU reac-tors, with almost no mathematics or equations. This monograph is also ideal as a reference for CANDU physicists, operators, regulatory staff, and for those who need to interact with reactor physicists at CANDU sites, nuclear laboratories, institutes, universities, or engineering companies. This monograph assumes prior knowledge of nucle-ar physics offered at the high-school level and/or at universities. Readers whose focus is only on calculations or on the development of software on reactor physics should refer to the books listed in the bibliography"-- Provided by publisher.

Identifiers: LCCN 2021016573 | ISBN 9780791884836 (hardback)

Subjects: LCSH: CANDU reactors. | Nuclear engineering--Canada. | Nuclear physics.
Classification: LCC TK9203.H4 S54 2021 | DDC 621.48/34--dc23

LC record available at https://lccn.loc.gov/2021016573

Series Editors' Preface

Nuclear Engineering and Technology for the 21st Century—
   Monographs Series

Nuclear engineering and technology play a vital role in achieving low carbon emission goals worldwide, while providing reliable, baseload power to the world economy. Presently over 10 percent of the world's energy needs are satisfied by nuclear power—with 30 countries operating 442 nuclear power plants and 3 countries (France, Slovakia, and Belgium) using nuclear power to provide over half their power needs (source: Nuclear Energy Institute: http://www.nei.org).

The country with the largest number of operational nuclear power plants (the United States) has 97 plants and uses nuclear power to provide over 19 percent of its needs. Concurrently, the advanced nuclear power plant designs are the basis for extensive, ongoing research and development efforts in many countries with the promise of enhanced sustainability, safety, and proliferation-free power-sources with everhigher operational efficiencies and capacity factors. Consequently, there are many fruitful topics of interest in the nuclear engineering field to be addressed in this exciting monograph series.

The *Nuclear Engineering and Technology for the 21st Century* monograph series provides current and future engineers, researchers, technicians and other professionals and practitioners with practical, concise but key information concerning the nuclear technologies from areas of medical applications, mining, processing and manufacturing, environmental monitoring to safe and energy-efficient plant operation and electricity generation. Each monograph should provide a well rounded and definitive state-of-the-art review of its subject, with a focus on applied research and development, and best industry practices, processes and related technological applications. The series is envisaged as a collection of 80 to 100 pages monograph publications which can stand as the most authoritative source of information on current state of a topic, application or discipline. Core topics include, but are not limited to:

+ best practices in power plant operation
+ nuclear science and technology in medicine
+ irradiation technologies and applications
+ fuel cycle processes, engineering and technologies
+ nuclear reactor thermal hydraulics and/or neutronics
+ materials for current and advance power generation
+ nuclear safety and environmental impact
+ next generation of nuclear power plants
+ radiation in our environment
+ radioecology, radiobiology, radiation chemistry

Series Editors:

Dr. Jovica Riznic, Canadian Nuclear Safety Commission
Dr. Richard Schultz, Idaho National Laboratory

Contents

List of Tables

List of Figures

Acronyms and Abbreviations

Some of the acronyms used in this report are given here, for convenience.

2D:	Two Dimensional
37M:	Modified 37-element
3D:	Three Dimensional
ACR:	Advanced CANDU Reactor
CANDU:	Canada Deuterium Uranium
CCP:	Critical Channel Power
CD:	Coolant Density
CHF:	Critical Heat Flux
CP:	Coolant Purity
CPPF:	Channel-Power Peaking Factor
CR:	Count Rate
CSEWG:	Cross Section Evaluation Working Group
CT:	Coolant Temperature
CVR:	Coolant-Void Reactivity
DCC:	Digital-Control Computers
DBA:	Design-Basis Accident
ENDF/B:	Evaluated Nuclear Data File/version B
FINCH:	Fully INstrumented CHannel
FPD:	Full-Power Day
FT:	Fuel Temperature
GSS:	Guaranteed Shutdown State
HBLP:	History-Based Local Parameter (Method)
HZP:	Hot Zero Power
HFDA:	Horizontal Flux Detector Assemblies
HFP:	Hot Full Power
ICFD:	In-Core Flux Detectors
IST:	Industry Standard Toolset
LLOCA:	Large Loss-of-Coolant Accident
LOCA:	Loss-of-Coolant Accident
LOF:	Loss Of Flow
LOMI:	Loss Of Moderator Inventory
LOR:	Loss of Regulation

LWR: Light-Water Reactor
LZC: Liquid Zone Controller
MB: Moderator Boron
MCA: Mechanical Control Absorber
MCNP: Monte Carlo N-Particle
MD: Moderator Density
mfp: mean free path
MGD: Moderator Gadolinium
MP: Moderator Purity
MT: Moderator Temperature
MTD: Margin To Dryout
MTT: Margin To Trip
NOP: Neutron Overpower Protection (same as ROP)
OID: Onset of Intermittent (fuel-sheath) Dryout
PCR: Power Coefficient of Reactivity
PWR: Pressurized-Water Reactor
PHTS: Primary Heat-Transport System
RFSP: Reactor Fuelling Simulation Program
RIH: Reactor Inlet Header
ROH: Reactor Outlet Header
ROP: Regional Overpower Protection (same as NOP)
RRS: Reactor Regulating System
SCM: Simple Cell Model
SDS: Shutdown System
SDS1:   Shutdown System One
SDS2:   Shutdown System Two
SG: Steam Generator
SLB: Steam-Line Break
SLOCA: Small LOCA
SOR: Shutoff Rod
SORO: Simulation of Reactor Operation
SPND: Self-Powered Neutron Detectors
TBD: Technical Basis Document
TSP: Trip SetPoint
VFDA: Vertical Flux Detector Assemblies
WIMS: Winfrith Improved Multigroup Scheme

Preface

Reactor physics is the physics of fission reactions and how they govern the behaviour of nuclear reactors. Arguably, its origins are linked to the middle of the twentieth century and the Manhattan Project. After World War II and the introduction of nuclear reactors as a source for generating electricity, the discipline went through the rapid development to become a mature and cornerstone of nuclear engineering. Several legacy textbooks are still in use in both undergraduate and graduate nuclear engineering programs worldwide. If this is so, is there really a need for another book on nuclear reactor physics? Well, we have to note that most of those highly valued legacy textbooks were written 20 or more years ago and there is no discussion on progress being made. On another note, nuclear and reactor physics were traditionally treated as scientific discipline heavily relying on sophisticated mathematical and computational methods and tools. As such, those textbooks were a must read and master for those professionals involved with reactor physics on a daily basis. However, I am confident that today's generation of practicing technicians and engineers working in various areas of research and development and supporting safe operation of nuclear reactors, whether in nuclear power plants, nuclear marine or research reactors, would greatly benefit from a book focussed on the physical phenomena and principles rather than on mathematical and numerical methods.

It turns out that the monograph in front of your eyes meets those needs perfectly. Wei Shen and Benjamin Rouben wrote a practical book focussed on physical phenomena and their engineering applications to ensure safe operation of nuclear power plants. Thus, it is my firm belief that the book will be a useful reference not only to reactor physicists but also for a much larger number of professionals working in operation, maintenance and regulation of nuclear facilities. I hope you will enjoy reading of this monograph.

Jovica Riznic, PhD., P.Eng., FASME
Editor-in-chief, *Nuclear Engineering and Technology for the 21st Century—Monographs Series*

1. Scope

From the educational point of view, there are many textbooks on reactor physics used at various universities in the world. However, most of these textbooks focus either on application to Light Water Reactors (LWRs), or on the theory and mathematics, with a significant number of equations and computational schemes. Or else they were written more than 20, or even more than 60, years ago, and therefore they do not reflect the evolution of reactor concepts and engineering requirements since then. All those categories of books are either difficult to follow for non-physicists working in the nuclear industry, or else are of little value for those who are interested in special features of CANDU reactor physics.

With more than 75 person-years of working experience in the area of CANDU and LWR reactor physics and safety, the intention of the authors of this monograph was to write a practical book on reactor physics for CANDU reactors, particularly for the safe operation of aged CANDU reactors, with almost no mathematics or equations! This monograph gives a glimpse of first principles and their engineering application in reactor physics, for those who are interested in or are working in the CANDU industry. This monograph is also ideal as a reference for CANDU physicists, operators, regulatory staff, and for those who need to interact with reactor physicists at CANDU sites, nuclear laboratories, institutes, universities, or engineering companies.

This monograph assumes prior knowledge of nuclear physics offered at the high-school level and/or at universities. As very few equations appear in the monograph, it is not considered suitable for specialists whose focus is only on calculations or on the development of software on reactor physics. Such readers should refer to the books listed in the bibliography at the end of the monograph.

Acknowledgements

Much of the information and many of the discussions and explanations presented in this monograph reflect what the authors have learned from material originally presented and explained in a great number of books, papers, and training documents. The authors would like to acknowledge these important sources of state-of-the-art knowledge on reactor physics, especially those on CANDU reactor physics.

1. D. Rozon, "Nuclear Reactor Kinetics", Polytechnic International Press, Montréal, Canada, 1998.
2. B. Rouben, "Introduction to Reactor Physics", Training Material, Atomic Energy of Canada Limited, Canada, 2002 September. All content except tables and figures is available from: https://canteach.candu.org/Content%20Library/20040501.pdf
3. "Science and Reactor Fundamentals Training Course", Volume 2, Rev. 1, Canadian Nuclear Safety Commission, Canada, 2003 January. https://canteach.candu.org/Content%20Library/20030101.pdf
4. "CANDU 6 Technical Summary", CANDU 6 Program Team, Atomic Energy of Canada Limited, Canada, 2005 May.
5. P. Reuss, "Neutron Physics", EDP Science, France, 2008
6. D.G. Cacuci, "Handbook of Nuclear Engineering", Springer Science & Business Media, Berlin, Germany, 2010.
7. "The Essential CANDU", Editor W. Garland, UNENE, Canada, 2014. http://www.unene.ca/education/candu-textbook
8. A. Hébert, "Applied Reactor Physics", Third Edition, Polytechnic International Press, Montréal, Canada, 2020.

2. Basics of CANDU Reactor Physics

2.1 Fission, Energy Released by Fission

2.1.1 Fission

Nuclear fission is the splitting of a (large) nucleus, with the release of energy. The nuclei of some heavy elements, such as U-238, do exhibit spontaneous fission in nature. However, the rate of such fissions is extremely low. The half-life of uranium is longer than 100 million years, and most of its decay is by alpha emission, so spontaneous fission is not a practical source of energy. Spontaneous fission is not of much use to us as an energy source!

To apply nuclear energy, we need to have a nuclear reaction which produces energy (by mass conversion), and which can be self-sustainable and controllable. Neutron-induced fission is such a reaction that can be used as a practical source of heat energy, from which we make steam to turn a turbine/generator.

The fission process occurs when the nucleus which absorbs the neutron is excited into an "elongated" (barbell) shape, with roughly half the nucleons in each part. This excitation works against the strong force between the nucleons, which tends to bring the nucleus back to a spherical shape, hence there is a "fission barrier" of height ~ 6 MeV. If the energy of excitation is larger than the fission barrier, the two parts of the barbell have the potential to completely separate, leading to binary fission!

The crucial feature of neutron-induced fission is that while one neutron enters the fission reaction, 2 or 3 or even more neutrons (about 2.4 on the average) are released in fission, and may be able to induce more fissions in other fuel nuclei. Fission neutrons make possible a self-sustaining chain reaction, in which the fission neutrons from one generation of fissions cause the next generation of fissions. The sustainability of the chain reaction depends on the fate of the neutrons. The chain reaction can be self-perpetuating ("critical") if at least one of the neutrons released in a fission is able to induce another fission. By judicious design, research and power reactors can be designed for self-sustainability (criticality), which is the crucial concept to be considered. We will do this further below. Controllability of the chain reaction is also extremely important, of course, to ensure that there is no "run-away" reaction. The total energy release is open to control by controlling the number of fissions. This is the operating principle of fission reactors.

Only a few nuclides can fission. A nuclide which can be induced to fission by an incoming neutron of any energy is called fissile. There is only one naturally occurring fissile nuclide: U-235. None of the other fissile nuclides, such as U-233, Pu-239 and Pu-241, are present in nature to any appreciable extent, but they can be (and are) produced in nuclear reactors. Fissionable nuclides such as U-238 and Pu-240 refer to nuclides that can be induced to fission, but only by neutrons of energy higher than a certain threshold. In summary, fissile nuclides are a subset of fissionable nuclides, and they are the easiest to fission. As a result, they provide the large majority of fissions in thermal reactors.

As the only naturally occurring fissile nuclide, U-235 is the most important nuclide for fission reactors. However, plutonium is also very important, because it is created in the reactor and it then participates in the chain reaction. Pu-239 is produced from the absorption of neutrons by U-238. By further neutron absorption in Pu-239, some Pu-240 and Pu-241 are also produced. Because U-238 yields fissile material via neutron absorption, it is called fertile.

2.1.2 Energy Released by Fission

The fuel nucleus splits into 2 smaller nuclei – the fission fragments (fission products), most of which are radioactive. A small fraction of the nuclear mass is turned into energy ($E = mc^2$) — close to ~200 MeV energy release per fission. The energy released during fission and the relative proportions of its components tend to be the same to within a few percent for all fissile nuclei. The sample breakdown given below refers to the energy released on thermal-neutron-induced fission of U-235 (all energies in MeV):

- Fission fragments: ~165 to 169
- Neutrons: ~5

- Prompt gamma photons: ~6 to 8
- Delayed gamma photons: ~6 to 7
- Delayed beta radiation (electrons): ~7 to 8
- Antineutrinos accompanying beta emission: ~8 to 12
  Total energy released: ~200 to 203

The kinetic energy of the fission fragments accounts for most of the energy released in the fission reaction:

- Over 80% of this energy is taken away by the two fragments and deposited at the point where the fission takes place within the fuel (for example, the mean free path (mfp) of the fission fragments in metallic uranium is only 7 μm).
- About 15% of the energy from fission appears as the kinetic energy of the neutrons and of the beta and gamma radiation. The energy of the electrons (betas) is deposited over a short distance in the fuel.
- The mfp of gamma photons is of the order of several centimetres. The gamma energy is mainly deposited in heavy materials: here again in the fuel, but over a wider area.
- The energy of the neutrons, mostly prompt neutrons, is mostly deposited in the moderator, which decelerates them.
- The energy of the antineutrinos, which interact extremely little with matter, is not recovered.

Thus, when calculating the total energy actually recovered, the "antineutrinos" line should be left out. However the energy produced by the radiative capture of the excess fission neutrons should be added (8 to 12 MeV). The final total recovered energy for uranium 235 would be about 200 to 203 MeV. This would be the fission energy used to calculate the fuel burnup (the concept of fuel burnup will be defined in Section 4.1). Three other examples of the energy released in fission of other nuclides are given for comparison:

- Uranium-238 (fast-neutron induced): ~205
- Plutonium-239 (thermal-neutron induced): ~210
- Plutonium-241 (thermal-neutron induced): ~212

2.2   Fission Products

The fission products are nuclides of roughly half the mass of uranium. However, note that they are not always the same in every fission. There are a great number of different fission-product nuclides, each produced in a certain percentage of the fissions. They are likely to have mass numbers between 70 and 160, with mass numbers near 95 and 140 the most probable. Symmetrical fission (equal-mass fragments) is rare.

Most fission-product nuclides are neutron rich; they decay typically by beta- or gamma-disintegration, and are therefore a potential radiological hazard, with various half-lives that range from fractions of a second to about 30 years. The ceramic uranium-fuel material (UO_2) and fuel sheath must encase the fission products to prevent them entering the Primary Heat-Transport System (PHTS) and leaving the reactor core. Fission products in the PHTS (or elsewhere outside the core) are a radiation hazard that would prevent equipment access even with the reactor shut down. Heavy shielding is required around the reactor for protection during operation, against the prompt radiation (neutrons and gamma rays). The shielding also limits exposure to gamma radiation that continues to be emitted by the fission products after shutdown. Fuel must be replaced remotely; special precautions must be taken in handling and storing used fuel.

Some of the fission products have high neutron-absorption cross sections and thereby "poison" the chain reaction in the reactor. A relatively high percentage of fissions produce the two most important poisons, Xe-135 and Sm-149, and these fission products capture a significant number of neutrons. Section 7 examines the effects of fission-product poisons.

Because many fission products are still decaying long after the fission reaction from which they originate, energy (heat) from this nuclear decay is actually produced in the reactor for many hours, days, even months after the chain reaction is stopped. This decay heat is not negligible. When the reactor is in steady operation, decay heat represents about 6-7% of the total heat generated. Even after reactor shutdown, decay heat must be dissipated safely, otherwise the fuel and reactor core can seriously overheat (to be discussed in Section 10).

2.3 Delayed Neutrons from Fission-Product Decay

Most of the fission neutrons emerge instantaneously from the reaction. However, a small fraction (about 0.5-0.6%) of the neutrons are produced by the decay of neutron emitters, daughters of some fission products (delayed-neutron precursors), and are thus "delayed". The delay corresponds to the decay time constant of the specific delayed-neutron precursor, ranging from a fraction of a second to minutes. In spite of their relatively small numbers, delayed neutrons play a crucial role in reactor regulation since they ensure the controllability of the sustained chain reaction (this will be discussed in Section 3).

The Chart of Nuclides identifies nearly one hundred delayed-neutron precursors. For thermal fission of U-235, the delayed neutrons are about 0.65% of all fission neutrons. This fraction, represented by the symbol β, is called the delayed-neutron fraction. For Pu-239, the delayed-neutron fraction is even smaller (about 0.23%). The value of β depends upon the actual nuclear fuel used. The weighted average depends on the particular mix of fuel nuclides in the reactor, a mix that changes with fuel burnup. For equilibrium CANDU fuel, the weighted-average delayed-neutron fraction β is close to 0.5%. Equilibrium fuel accounts for fissions of all fissile nuclides in the core, as well as fast fission of U-238.

The delayed-neutron precursors can be grouped according to their half-lives, ranging from fractions of a second to about 60 s. Table 1 lists the fractional neutron yields for each delayed-neutron-precursor group for U-235. It is customary to categorize delayed neutrons into 6 delayed-neutron groups, but fewer or more groups can be used. For kinetics analysis of time frames of the order of 5 seconds, 6 delayed-neutron groups generally provide sufficient accuracy; for longer time frames, a greater number of groups might be needed.

Delayed neutrons do not have the same properties as prompt neutrons released directly from fission. The average energy of prompt neutrons is about 2 MeV. This is much greater than the average energy of delayed neutrons (about 0.5 MeV). The fact that delayed neutrons are born at lower energies has two significant impacts on the way they proceed through the neutron life cycle. First, delayed neutrons have a

Table 1 Delayed-Neutron Data for Thermal Fission in U-235
(*Source:* Table 3.5 from [1])

Group	Half-Life (sec)	Decay Constant (l_i, sec^{-1})	Energy (ke V)	Yield, Neutrons per Fission	Fraction (β_i)
1	55.72	0.0124	250	0.00052	0.000215
2	22.72	0.0305	560	0.00346	0.001424
3	6.22	0.111	405	0.00310	0.001274
4	2.30	0.301	450	0.00624	0.002568
5	0.610	1.14		0.00182	0.000748
6	0.230	3.01		0.00066	0.000273

Total yield: 0.0158
Total delayed fraction (β): 0.0065

much lower probability of causing fast fission than prompt neutrons, because their average energy is lower than the minimum required for fast fission to occur. Second, delayed neutrons have a lower probability of leaking out of the core, because they are born at lower energies and subsequently travel a shorter distance than fast neutrons. These two considerations (smaller fast-fission factor and higher fast-non-leakage probability for delayed neutrons) are taken into account by a term called the importance factor. The importance factor relates the weighted-average delayed-neutron fraction to the effective delayed-neutron fraction.

The effective delayed-neutron fraction is defined as the fraction of neutrons at thermal energies which were born delayed. The effective delayed-neutron fraction is the product of the average delayed-neutron fraction and the importance factor. In a small reactor with highly enriched fuel, the increase in fast-non-leakage probability will dominate the decrease in the fast-fission factor, and the importance factor will be greater than one. In a large reactor with low enriched fuel, the decrease in the fast-fission factor will dominate the increase in the fast non-leakage probability and the importance factor will be less than one; it is about 0.97 for commercial LWRs. For conservatism, such an importance factor should be applied to CANDU as well, for instance by using a scaling factor on β.

2.4   Delayed Photoneutrons

Besides "direct" delayed neutrons, i.e., those born "in the fuel", there is another group of delayed neutrons in CANDU reactors - the delayed photoneutrons. Delayed photoneutrons are born in the heavy water (mostly in the moderator) when a high-energy gamma ray (energy > 2.224 MeV, the binding energy of D - deuterium, heavy hydrogen) from the gamma-decay of certain fission products breaks up a D into a proton and a neutron. There are several precursors in irradiated fuel which can emit such energetic gammas. In addition, high-energy gammas can be emitted from some nuclides produced by the long-term irradiation of the calandria vessel and in-core structural materials (such as pressure tubes and calandria tubes) in CANDU reactors.

The term effective photoneutron precursor concentration is used because the effective concentrations must also account for the fact that not all emitted gamma rays will result in the production of a photoneutron and that the fraction of photoneutrons emitted depends on the geometrical arrangement of the core lattice. Photoneutron precursors can be grouped by their decay constant, similarly to the precursors of delayed neutrons from the fuel. It is customary to use 11 delayed-photoneutron groups, for a total of 17 delayed-neutron groups. Once the delayed-photoneutron groups have been defined, photoneutrons are treated no differently from "regular" delayed neutrons in the neutron-kinetics analysis.

In the critical reactor, the delayed-photoneutron source is considerably weaker than the delayed-neutron source — the delayed-photoneutron fraction is about 0.03% of all fission neutrons. However, the half-lives of the delayed-photoneutron precursors are generally much longer (up to 10^6 s) than the half-lives of the delayed-neutron precursors, as shown in Table 2. Hence, delayed photoneutrons are important in low-power operation and reactor startup (to be discussed in Section 10). The delayed-neutron data is used to estimate neutron power during outages, to gauge whether ion chambers will be on-scale and whether startup instrumentation may be needed for monitoring the reactor power during reactor restart (to be discussed in Section 5.4).

Table 2 Group Yields and Half-Lives for 11 Groups of Delayed-Photoneutron Precursors (*Source:* Table 2 from [2])

Group #	Fractional Group Yield	Half Life
1	0.0011	307.6 h
2	0.0023	53.0 h
3	0.0073	4.4 h
4	0.0527	5924.0 s
5	0.0466	1620 s
6	0.0757	462.1 s
7	0.1576	144.1 s
8	0.0448	55.7 s
9	0.2239	22.7 s
10	0.1940	6.22 s
11	0.1940	2.3 s

In those 17 groups of delayed-neutron data, the longest half-life is 12.8 days, that of Ba-140 contributing to delayed-photoneutron group one. While sufficient for short outages, the sets of 11-group delayed-photoneutron fractions and time constants shown in Table 2 do not provide an accurate prediction of neutron power for long outages (greater than two and a half months). In an effort to overcome this deficiency, many CANDU utilities have created additional delayed fractions and groups by fitting to historical long-outage measurements.

2.5 Neutron Moderation

Neutrons that are born in fission have high energy, a distribution of energies that has a maximum at ~1 MeV (neutron speed ~13,800 km/s). However, fast neutrons produced by fission are less likely to cause fission than thermal neutrons. In U-235, the fission cross section (the concept of cross section will be defined in Section 2.6.2) for thermal neutrons is 580 barns, compared to one or two barns for fast fission.

Fission neutrons tend to lose energy by collisions with materials in the reactor. However, the energy loss per collision is very small in collisions with uranium, because the uranium nuclei are so heavy. Before its kinetic energy is reduced to thermal values, the neutron will reach an energy range (the resonance-energy range) where the absorption cross section of U-238 is very high. In a collision with uranium in this energy range, the neutron will be absorbed but will not induce fission; this is called resonance capture. Since most of the uranium fuel consists of U-238 (more than 100 tonnes in a large CANDU, compared to less than 1 tonne of the fissile nuclides), the resonance-capture loss makes it impossible to achieve a chain reaction in natural uranium or low-enriched (<5 wt%, percentage by weight, of U-235) uranium fuel unless it is possible somehow to reduce the number of neutrons that are in the vicinity of uranium when their energy is in the resonance range. It is preferred to have neutrons proceed to lower energies.

Therefore, most reactors today use a "moderator" to slow down the neutrons, to increase the likelihood of perpetuating the fission chain reaction. To maximize the chance of fission, the moderator (a material with light atoms, of mass similar to that of a neutron) slows neutrons quickly through the resonance-energy range, without absorbing them once they are thermalized. The maximum extent to which neutrons can be slowed down is to energies at which the neutron population is in thermal equilibrium with the ambient environment ("thermal" energies). The process of slowing neutrons down this way is also called "neutron thermalization", and these neutrons are labelled "thermal". For a temperature of 20°C, thermal neutrons have a near-Maxwellian distribution of energies, with a most probable energy of ~0.025 eV, corresponding to a speed of 2.2 km/s, orders of magnitude lower than that of fast neutrons [although still approximately the speed of a bullet].

Reactors which rely very predominantly on fissions induced by thermal neutrons are called thermal reactors. CANDU is a thermal reactor. There are three possible moderators for use in thermal reactors:

- H in water: cheap but captures lots of neutrons. This is why LWRs must enrich the fuel.

- D in heavy water: almost as good as H in slowing down neutrons, and absorbs much fewer neutrons. The downside of heavy water is that it is much more expensive than water.
- C (graphite)

For a natural-uranium reactor using uranium dioxide fuel, the only possible moderator is heavy water. The choice of the relative volumes of fuel and moderator and their geometric arrangement determines the degree of sustainability of the chain reaction.

Notwithstanding the above discussion focusing on thermal reactors, note that some reactors are designed to work with fast neutrons, so have no moderator. But most commercial power reactors are thermal reactors.

2.6   Reaction-Related Quantities

To follow the fate of a neutron as it bounces randomly through a mixture of reactor materials we need a quantitative comparison of the possible reactions. Most convenient is a comparison of reaction rates, which depend on nuclear cross sections and neutron flux, quantities we will now introduce.

2.6.1 Neutron Flux and Neutron Current

The neutron flux (φ) measures the intensity of neutrons passing through a cubic centimeter of material. It is given by $\varphi = nv$, where n is the neutron density (number of neutrons per cm^3) and v is the speed of the neutrons (cm/s). The standard unit for flux is $neutrons.cm^{-2}s^{-1}$.

As neutron velocity has a direction, it is important to first consider and define the angular flux, which is a vector function of position, neutron energy, time, and neutron speed in its direction of motion. Once we have the angular flux in all directions, we can then integrate (sum up) the angular flux over all directions, to obtain the angle-integrated flux. This can be interpreted as the total distance travelled in one second by all the neutrons in one cm^3 volume at the given position and given time. Note that when the context is clear, the term "angle-integrated flux" is often shortened simply to "flux".

Flux is now the established term in reactor physics, but the quantity should really be called "flux density" instead, because this is not a flux in the usual sense of a quantity passing through a surface, but a flux density defined on the basis of the volumetric concept of density. Reactor physicists use the term "current" to denote what is known as "flux" in other branches of physics: the number of neutrons passing through a surface element, normalized per unit surface and unit time.

When the neutron flux density is not spatially uniform, there is at any point a net leakage of neutrons from regions of high flux density to regions of low flux density. The net movement of neutrons at any point can be expressed in terms of the neutron current, denoted by the vector quantity J. Just as the neutron flux depends on the position in the reactor, so does the neutron current. In neutron-diffusion theory, it can be shown that the net neutron current is approximately proportional to (the negative of) the gradient of the flux: $J = -D\nabla\varphi$, where the proportionality constant D is called the diffusion coefficient. D is a function of the properties of the medium, and also depends on the neutron energy (or speed).

2.6.2 Cross Sections

As neutrons diffuse through the materials of the reactor core, they may enter into a number of reactions with nuclei of various elements: scattering (elastic or inelastic, depending on whether the energy state of the nucleus is changed), absorption, fission, or other reaction. Imagine a monoenergetic beam of neutrons impinging upon a (very thin) slice of a target material. We may think of the nuclei in the target as presenting a certain effective area to incoming neutrons for a given type of reaction (e.g. scattering, absorption, fission, etc.). This area is called the microscopic cross section σ, the effective area presented to the neutron by one nucleus of the material.

It is important to note that σ represents the reaction probability with a particular nucleus. In general, σ depends on the type of nucleus, the type of reaction studied, and the incoming neutron's energy. It has units of area and can be expressed in cm^2, although a much more appropriate unit is the barn = 10^{-24} cm^2, or sometimes the kilobarn (kb) = 10^{-21} cm^2.

The rate at which a reaction takes place in bulk material depends not only on the reaction probability with a particular nucleus (represented by the microscopic cross section) but also on the density of target nuclei. To account for this we introduce a quantity known as the macroscopic cross section Σ, which is the product of the microscopic cross section (σ) of an individual nucleus of a given material and the atom (or nuclei) number density, N (that is, the number of atoms per cm^3) of the material in the region considered. Thus the macroscopic cross section is given by: $\Sigma = N\sigma$. The unit for Σ is cm^{-1}.

If several different types of nuclei are present in the material, then a number of partial products $N\sigma$ for the various nuclide types must be added together to give Σ. The microscopic cross section σ is a basic physical quantity which is determined by experiments, where neutron beams of various energies are made to interact with target materials. Both σ and Σ depend on the material, the neutron energy or speed, and the type of reaction.

The inverse of the macroscopic cross section for a given material, which has the dimensions of a distance, does have an easily visualized meaning. For example, the quantity $1/\Sigma a$ equals the average distance that a neutron will travel before being absorbed by the material, and is known as the absorption mfp.

In practice, the materials placed in reactors are isotropic, which means that they have the same properties no matter which angle they are seen from. Consequently, cross sections are not dependent on the direction of the incident neutron, but only on its speed (or energy).

2.6.3 Reaction Rates

The point reaction rate, R, is the number of reactions per second per cubic centimetre of material. For a reactor type x, the mfp is $1/\Sigma_x$. The flux (φ) is the total neutron track length traversed by the neutrons in one second in one cm^3, so dividing flux by the length of track required (on average) for one reaction of type x, we get the total number of reaction of type x, that is: $R = \varphi\Sigma_x$. With φ in cm^{-2} s^{-1} and Σ_x in cm^{-1}, R has units $cm^{-3} \ s^{-1}$. In practice it is important to account for all the different types x of reaction, by considering the cross sections Σ_x and reaction rates R_x, labelled with their respective subscript x.

2.7   Neutron Life Cycle in CANDU

There is therefore a "neutron cycle" in a thermal reactor: fast neutrons are born in fission, they are slowed down in the moderator, they get absorbed in resonances or are thermalized or leak out. When they are thermalized, they provide the large majority (~95%) of further fissions. Figure 1 shows a CANDU-reactor's basic lattice cell. For reference, the CANDU lattice pitch (centre-to-centre distance between nearest-neighbour channels) is about 28.6 cm. Though this is not the optimum value in the sense of maximizing the lattice reactivity or "chain-reaction potential" (the optimum value is over 35 cm), the spectrum is highly thermal at this lattice pitch: almost 99% of the neutrons in the cell are in the thermal group (E < 0.625 eV). The fuel-bundle's diameter (or pressure-tube's inner diameter) is about 10 cm.

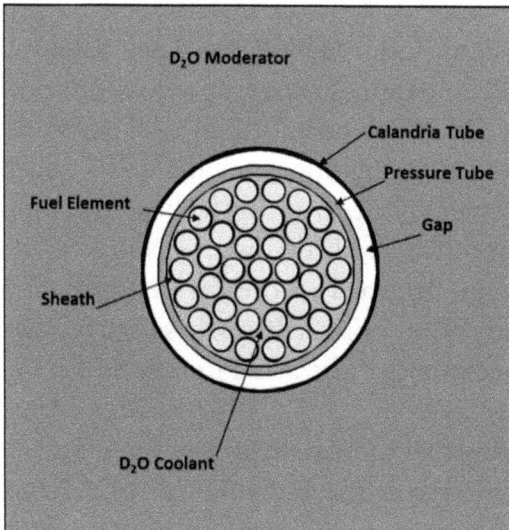

Figure 1   Face View of the CANDU Basic Lattice Cell

The fate of a neutron depends on the relative probabilities of the scattering and absorption reactions for the various nuclei it encounters as it diffuses through the CANDU lattice. These probabilities depend on the relative abundance of the different nuclei and the cross sections for particular reactions.

Elastic collisions with nuclei in the moderator slow a fast neutron until it reaches thermal energy. In a CANDU reactor, the average distance spanned by a neutron as it is being thermalized is about 25 cm and the time is a few microseconds. It takes about 36 collisions with deuterium to thermalize a fast neutron with a kinetic energy of 2 MeV to a thermal neutron with a kinetic energy of 0.025 eV $(= 0.6^{36} \times (2 \times 10^6))$, assuming 60% of energy loss by a neutron per collision.

Following thermalization, the neutron diffuses as a thermal neutron until some material, most probably fuel, absorbs it (unless it reaches the edge of the reactor and leaks out). The time between thermalization and absorption accounts for most of the neutron's lifetime, about one millisecond $(10^{-3}$ s). The average distance between birth and absorption is almost 40 cm (close to the diagonal distance across the CANDU lattice).

In CANDU, about 95% of the overall neutron production comes from thermal fissions (induced by thermal neutrons in U-235, Pu-239, and Pu-241). The remaining neutron production comes from fast fission (induced by fast neutrons in U-238). All neutrons born by fission are fast neutrons. A careful choice of materials and their configuration keeps the net loss of neutrons by leakage and capture in non-fuel materials to about 20% of all the neutrons produced by fission. The remaining 80% are absorbed in the fuel as thermal neutrons. With our choice of fuel, only about half of these cause fission, just enough to return the same number as started the cycle ($\sim$2.5 neutrons/fission $\times \sim$40% = 100%).

It is worth noting that more than 40% of the energy is produced from plutonium over the life of the fuel in a CANDU reactor. Actually, in fuel near the exit burnup, plutonium contributes about three-quarters of the fission energy generated in this fuel.

2.8   Fundamental Features of the CANDU Fuel Lattice

The following three fundamental features have "defined" the CANDU fuel-lattice design from the start:

- Heavy-water moderator and heavy-water coolant
- Natural-uranium fuel bundle
- Horizontal pressure tube for on-power refuelling

2.8.1 Heavy-Water Moderator and Heavy-Water Coolant

The CANDU reactor design has opted for heavy water as the moderator, to take advantage of the neutron economy provided by deuterium. This allows the use of natural-uranium fuel and precludes the need for expensive fuel-enrichment technology.

The first desirable property for a moderator is the ability to thermalize neutrons in as few collisions as possible. It is also important that a moderator have a small probability of capturing neutrons (i.e., a small neutron-absorption cross section), otherwise there will be a negative impact on neutron economy. In this respect hydrogen is not the best moderator, because it has a relatively high neutron-absorption cross section. The best (although expensive) moderator for neutron economy is heavy water (D_2O), which is why it was chosen as the moderator for CANDU. The fission chain reaction cannot be self-sustaining with light water (H_2O) as moderator and natural uranium as fuel.

Heavy water is used as coolant instead of light water in all operating CANDU reactors for additional neutron economy. It circulates in the pressure tubes in the reactor and takes away the heat produced in the fuel. This coolant system is also called the PHTS. The typical coolant temperature within a fuel channel increases from ~266°C at the channel inlet to ~312°C at the channel outlet. The PHTS heavy-water inventory is around 150 tonnes, and the coolant flow is ~1900 kg/s per pass (for the CANDU 6). The coolant leaving the core can contain a certain void fraction due to boiling at the exit of some of the high-power channels.

The moderator and the coolant are separated in the CANDU design. It is important to note the benefits and the safety advantage of having a moderator much cooler (at ~70°C) than the coolant. Consider the following aspects:

- The relatively cool moderator provides a large and important heat sink in the case of a loss-of-coolant accident (LOCA) and of a severe accident with impaired heat-sink capability.
- Reactivity devices are located in the moderator, interstitially between pressure tubes, thus in a benign (low-temperature and low-pressure) environment — removing the possibility of pressure-assisted rod ejection, a distinctive safety advantage.
- The moderator provides cooling to any reactivity devices which may absorb heat;
- The thermalization efficiency of the moderator depends on its density; a cold moderator is more efficient in slowing down fast neutrons.

Because of the significant probability of neutron capture by any light water present in the heavy-water moderator, it is crucial that the latter have very high isotopic purity. Reactor-grade heavy-water moderator is at least 99.75 wt% D_2O. The moderator efficiency increases sharply when the purity increases (see Section 8.7). Even a reduction of 0.1% in the isotopic purity has a significant negative effect on the neutron economy of the reactor and on the achievable fuel burnup. The isotopic composition of the moderator should be monitored and kept as high as possible.

One by-product of heavy water is the generation of delayed photo-neutrons which add ~ 6% to the delayed-neutron fraction,

2.8.2 Natural-Uranium Fuel Bundle

The main difference between the fuel-management strategies of CANDU reactors and LWRs is the type of nuclear fuel which is used: natural uranium in CANDU reactors versus enriched uranium in LWRs. The reason for this choice is that Canada has substantial, high-grade, domestic uranium mineral resources and, for a reactor cooled and moderated by heavy water (D_2O), it is not necessary to use enriched uranium fuel. This is very convenient for countries which wish not to have to rely on expensive, and most probably foreign, enrichment technology.

However, the CANDU design is very flexible and allows the use of advanced fuel cycles, using slightly enriched uranium (SEU), recovered uranium (RU), mixed-oxide fuel (MOX), thorium fuels (Th), natural-uranium equivalent (NUE), Direct Use of spent PWR In CANDU (DUPIC), and actinide burning, etc. These can be introduced into CANDU with few or no hardware changes, when the option becomes attractive.

CANDU fuel is of very simple design. It is manufactured in the form of elements of length ~48 cm. Each element consists of uranium-dioxide pellets encased in a zircaloy sheath. A number of fuel elements are assembled together to form a bundle of length ~50 cm. The elements are held together by bundle end plates. The CANDU fuel bundle contains only 7 different components and is short, easy to handle, and economical.

Two bundle types are used in operating CANDU reactors: the 28-element bundle (in Pickering) and the 37-element bundle (in Bruce, Darlington and the CANDU 6). The 28-element bundle has a smaller total surface area of fuel and a smaller ratio of sheath mass to fuel mass than the 37-element bundle, which gives the 28-element bundle a reactivity advantage. On the other hand, the 37-element bundle features better thermalhydraulic properties due to the greater fuel subdivision, as the larger number of pins of smaller diameter provide a better heat-removal capability. Thus, the 37-element bundle can operate at a higher power than the 28-element bundle. This tends to further reduce the reactivity of the 37-element bundle, but allows a higher total reactor power for the same mass of fuel, an important economic advantage from the point of view of a plant's capital cost ($ per MWe installed). In the 2010s, the modified 37-element (37M) fuel bundle with reduced centre-pin diameter was developed and is currently used in most operating CANDU units (in Bruce, Darlington and some CANDU 6 units) to compensate for the reduced safety margin caused by pressure-tube creep in the aged CANDU units.

The natural-uranium fuel has much lower excess reactivity than that of the enriched fuel. CANDU natural-uranium fuel achieves a discharge burnup in the ~7,200–9,500 MWd/Mg(U) range, depending on the specific reactor design. This is much lower than the discharge burnup in LWRs ($\sim$30,000–50,000 MWd/Mg(U)). However, considering fuel enrichment for LWRs (about 3.5%–5% U-235) and the depleted-uranium tails which are discarded, CANDU reactors are more efficient than LWRs and extract $\sim$25% more energy per Mg of mined natural uranium.

With natural uranium as the nuclear fuel, on-line refuelling (will be descibed in Section 4.3) becomes practically mandatory to provide enough reactivity to allow continuous full-power production. The horizontal pressure-tube design has been adopted in CANDU as the best option for the refuelling of individual fuel channels.

2.8.3 Horizontal Pressure-Tube Design

A major characteristic, selected early in the development of the CANDU reactor, is the pressure-tube design. The PHTS pressure in CANDU is $\sim$100 atmospheres. The pressure tubes are made of an alloy of zirconium and 2.5% niobium. The pressure-tube concept was originally chosen for CANDU because the manufacture of a pressure vessel of the size required for a CANDU would at the time have challenged the capability of Canadian industry. The pressure-tube design has both advantages and disadvantages in relation to the design and safety of the reactor as outlined below:

- One consequence of the pressure-tube design is the physical separation of the moderator from the coolant. The coolant and moderator do not have to be identical and could be of different nature altogether.
- The purpose of the calandria tube is to prevent the moderator from contacting the very hot pressure tube, which would cause the moderator to boil. The insulating gas gap between the pressure and calandria tubes allows the moderator to be at low temperature ($\sim$70°C) and at close to atmospheric pressure.
- The pressure-tube concept allows the regular replacement of fuel in the reactor while on power, precluding the need for periodic shutdowns for refuelling.
- Horizontal pressure tubes promote symmetry because coolant can be circulated in opposite directions in alternate tubes (i.e., using bi-

directional coolant flow), making average nuclear properties essentially identical at the two ends of the reactor (unlike the situation in LWRs).

- Horizontal pressure tubes also facilitate bi-directional refuelling (i.e., refuelling in opposite directions in adjacent channels), which further promotes symmetry.

- The sudden rupture of a pressure tube would not be a catastrophic event as could be the rupture of a pressure vessel.

- The entire core structure, including pressure tubes and calandria tubes, is under maximum gamma and neutron irradiation. Under irradiation and on account of neutron absorption by the zircaloy, the pressure tubes creep and provision must be made to accommodate the pressure-tube expansion. When necessary, the pressure tubes must be replaced to reduce the mechanical stresses on the calandria walls and extend the life of the reactor.

- The pressure tube has to support a channel full of fuel and heavy water along the entire core length without intermediate core support. There are only 4 zircaloy "garter-spring spacers" in the annulus gap between pressure tube and calandria tube, to prevent contact between the two tubes. The weight of the fuel channel (full of fuel and heavy water) will result over time in pressure-tube sagging.

- Whereas the loss of coolant shuts an LWR down, the loss of coolant in a CANDU reactor does not have the same effect, because the coolant is separated from the moderator. Instead, a LOCA in CANDU results in a power pulse, which must be quickly turned around by shutdown-system action to avoid overheating the fuel.

- On-power refuelling requires frequent opening of the PHTS pressure boundary. At that time, it is possible that a small LOCA or a channel flow blockage could occur.

- Each channel refuelling creates a small localized perturbation in the overall power distribution, that has to be evaluated before and after each refuelling. In order to calibrate the theoretical calculations, a large number of detectors have to be installed, monitored and calibrated.

3. Neutron Kinetics and Reactor Control

3.1 Multiplication Factor, Reactivity, and Concept of Criticality

A nuclear reactor is designed to achieve the very delicate balance between neutron "production" (release) in fission reactions and neutron loss by absorption and leakage. A given neutron will be "born" in a fission event and will then usually scatter about the reactor until it meets its eventual "death" either by being absorbed in some material or by leaking out of the reactor. A certain number of these neutrons will be absorbed by fissionable nuclei and induce further fissions, thereby leading to the birth of new fission neutrons, that is, to a new generation of neutrons. The ratio of the number of neutrons born in a fission-neutron generation to the number born in the previous generation is called the effective reactor multiplication factor, k_{eff}. The k_{eff} characterizes the balance or imbalance in the chain reaction. Alternatively, k_{eff} can be defined by the ratio of production rate to loss rate of neutrons in the reactor. These definitions are given below:

$$k_{eff} = \frac{Rate\ of\ neutron\ production}{Rate\ of\ neutron\ loss\left(by\ absorption\ and\ leakage\right)}$$

or, in another definition,

$$= \frac{Number\ of\ neutrons\ born\ in\ one\ generation}{Number\ of\ neutrons\ born\ in\ previous\ generation}$$

Balance exists if $k_{eff} = 1$, and the number of neutrons will stay constant in time. Another useful quantity related to k_{eff} can be defined: the amount by which $(1/k_{eff})$ differs from unity defines reactivity (ρ):

$\rho = 1 - 1/k_{eff}$ = fractional change in neutron population from one generation to the next

The reactivity is another measure of the balance or imbalance in the chain reaction. Reactivity is also sometimes used in the sense of an increment, e.g., adding positive or negative reactivity to increase or decrease the system reactivity. The multiplication constant k_{eff} is a ratio of like quantities, hence it is a pure number and has no units. Similarly, the reactivity ρ is a pure number and has no units. However, because in reactor physics we seldom deal with k_{eff} values extremely different from unity, reactivity is often written in "units" of a small fraction of unity. Three of the common units are:

- 1 milli-k (or mk) = 0.001. ρ = 1 mk means neutron production > loss by 1 part in 1000
- 1 pcm = 10^{-5} = 0.01 mk
- 1 dollar = β_{eff} (effective delayed-neutron fraction)

There are three possibilities for k_{eff} or ρ:

- If $k_{eff} > 1$ ($\rho > 0$), the number of neutrons increases in every successive generation, i.e., the chain reaction is more than self-sustaining. The neutron population and the power level increase with time. The reactor is said to be supercritical.
- If $k_{eff} = 1$ ($\rho = 0$), the chain reaction is self-sustaining and the reactor is critical. The neutron population and power level are in steady state. In this situation there is an exact balance between neutron production and neutron loss.
- If $k_{eff} < 1$ ($\rho < 0$), the chain reaction is not self-sustaining. The reactor is said to be subcritical. In this case, the neutron population and the power level decrease with time except if there is an external source present in the reactor (more on this in Section 9.2: Subcritical Multiplication Equation).

Note that both k_{eff} and ρ relate to the balance or lack thereof in the chain reaction; they do not refer to the power level. Criticality does not mean that the reactor is at full power. The reactor can be critical at any power level: full power, very low power, power greater than normal. During startup, the reactor typically goes critical at a very small fraction ($\sim 10^{-6}$) of full power. In a critical reactor the power level is steady. On the other hand, departure from criticality signifies that the power level is changing: increasing when the reactor is supercritical, decreasing when it is subcritical.

3.2 Prompt Neutron Kinetics - Power Increase Without and With Delayed Neutrons

More than 99% of the neutrons in a reactor are "prompt" neutrons – they are born at the instant of fission. There are also "delayed" neutrons (see near the end of this section). For the moment we consider the (imaginary) situation in which there are only prompt neutrons. The neutron lifetime l_p refers to the time interval between prompt-neutron generations, which can be conveniently considered as comprising a slowing-down time ($\sim 10^{-7}$ s) and a neutron-diffusion time ($\sim 10^{-3}$ s in CANDU). For small reactivity insertions, the variation of the neutron population (and, consequently, of the power P) will depend on the neutron generation time Λ, the average time interval between successive neutron generations which would be the neutron lifetime l_p if there were no delayed neutrons. In a simplistic treatment of neutron kinetics, in fact, the power varies exponentially with reactivity insertion ρ and with time t in units of Λ:

$$P(t) = P_0 \exp(\rho t / \Lambda)$$

The reactor period T, defined as $T = \Lambda / \rho$, is the time it takes power to rise (or fall) by a factor e [the base for natural logarithms ($e \approx 2.718...$)]. Dividing the exponential relationship by P_0 and taking the natural logarithm gives the equivalent relationship:

$$ln\ [P/P_0] = t/T \text{ or } log\ rate = (lnP - lnP_0)/t = 1/T$$

This shows that the rate of change of the logarithm of the power (log rate in station jargon) is the inverse of the reactor period. The log rate represents a fractional increase per second. In CANDU control rooms, one can see "log-rate meters" that give the log rate in units of percent present power per second (% P·P· s^{-1}).

The equation above indicates that under a positive value of reactivity perturbation ρ, the neutron population grows exponentially with time and diverges for as long as the perturbation lasts. The parameter that characterizes the rate of power rise is the reactor period T, which is proportional to the average neutron generation time Λ, and inversely proportional to changes in reactivity ρ. A long reactor period corresponds to a slow power rise, a short period to a fast power rise.

As a numerical example, consider a 2-mk reactivity increase in a critical reactor, i.e., $k_{eff} \approx 1.002$. If there were no delayed neutrons, the average neutron generation time Λ would be the neutron lifetime l_p which would be as short as $\sim 10^{-3}$ s in CANDU. Then the reactor period $T = 0.001/0.002 = 0.5$ s, and after a second $P = P_0 e^{1/0.5} = 7.4 \cdot P_0$. If the departure from criticality is 10 times higher, ρ = 20 mk, then the reactor period will be 10 times shorter: $T = 0.001/0.02 = 0.05$, and the power will be increasing by a factor of 7.4 every 0.1 second! This is a very steep rate of power increase, which would render reactor control extremely difficult or impossible. In summary, T needs to be long to control the reactor. It is worth noting that Λ is about 30 times shorter in LWRs compared to that in CANDU reactors, and the rate of change of power would then be 30 times as great in LWRs for the same reactivity.

It is clear from this discussion that prompt-neutron kinetics poses safety concerns because there may be very small reactor periods. Safe reactor control would not be possible if the neutron population consisted only of prompt neutrons.

In reality, not all neutrons are born prompt, and a very small fraction (~0.5-0.6%) are delayed neutrons with an average neutron generation time of about 13 s. These delayed neutrons, in spite of the small delayed fraction, drastically increase the average neutron generation time, $\Lambda = .995 \times 0.001 + .005 \times 13 = 0.066$ s. Thus, for a 2-mk reactivity increase, $T = 0.066/0.002 = 33$, and after 1 second $P = P_0 e^{1/33} = 1.03 \cdot P_0$, after 5 seconds $P = P_0 e^{5/33} = 1.16 \cdot P_0$, and after 10 seconds $P = P_0 e^{10/33} = 1.35 \cdot P_0$, a much more manageable power increase. Hence, delayed neutrons are crucial for reactor control (see Figure 2).

3.3 Prompt Criticality

To this point, we have considered only small reactivity additions, ρ << β, or at least ρ values not very close to β. What happens if we insert a large positive step change in reactivity, greater than β?

Assuming the multiplication factor is k, N neutrons of one "generation" give rise to a total of kN in the next "generation". The majority of the

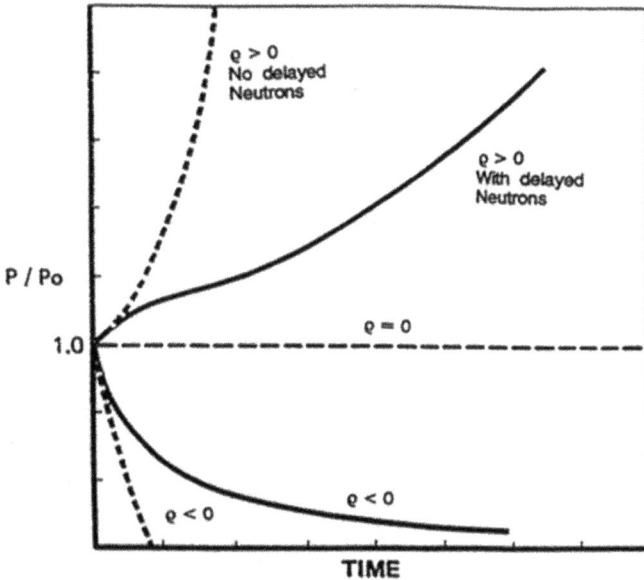

Figure 2   Effect of Delayed Neutrons on Power Transients (**Low-Reactivity Transient**)

neutrons are prompt, so the prompt fraction of neutrons, $(1 - \beta) \times kN$ neutrons, appear immediately [$(1- \beta)$ is called the prompt fraction]. The delayed neutrons, numbering $\beta \times kN$, are not born immediately but are instead released later, from the "bank" (concentration) of delayed-neutron precursors. The delayed neutrons feed into the cycle at a rate controlled by the delayed-neutron lifetime and the size of the precursor "bank".

A reactor with an effective multiplication factor large enough to make the prompt neutrons increase sufficiently to make the reactor reach criticality without waiting for the delayed neutrons is called "prompt critical". The condition for prompt criticality is $(1-\beta) \times kN = N$, i.e., $\beta = 1 - 1/k = \rho$. Depending on the values of ρ (the reactivity insertion associated with a given reactor transient), and of β (the delayed-neutron fraction), we can classify the different reactor states and the related consequences as follows:

- if $0 < \rho < \beta$, the reactor state is simply called supercritical. The divergence related to prompt kinetics is avoided. The power variation is then dominated by delayed neutrons.

- if $\rho \geq \beta$, the reactor state is called prompt critical or prompt super-critical. The fission chain reaction is entirely self-sustaining on prompt neutrons alone, without having to wait for the delayed neutrons. Under these circumstances, the exponential power rise from the prompt neutrons alone is similar to that in the example where we assumed that all neutrons were prompt.

3.4 Summary of Neutron Kinetics

In summary, unless prompt supercriticality is reached, the variation of the neutron population (and of the power) is not dominated by the prompt neutrons, but instead by the decay process of the precursors and by the emission of delayed neutrons. The presence of the delayed neutrons slows down considerably the rate of power changes in transients (note that this statement is true for negative reactivity insertions as well). This allows longer reactor periods, which enables the appropriate control of the neutron population within safe procedures.

In normal operation, reactors are controlled so they are critical on the sum of prompt and delayed neutrons. Two parameters that affect the rate of rise in power after a given positive reactivity insertion are the delayed-neutron fraction β and the prompt-neutron lifetime l_p. For small reactivity insertions, the rate of change of power will be similar for all power reactors. On the other hand, for large reactivity insertions (close to or above the delayed-neutron fraction β), the rate of change of power will be different for different reactors on account of differences in the delayed-neutron fraction β and the prompt-neutron lifetime l_p.

Figure 3 shows the reactor period T vs. reactivity insertion ρ for various prompt-neutron lifetimes l_p for different types of reactors. It is important to note that delayed neutrons play a moderating effect only if $\rho < \beta$. As reactivity ρ comes close to β, the reactor becomes less and less dependent on the delayed neutrons, and the reactor period T decreases smoothly to shorter and shorter values. Numerical analysis shows that the reactor period is less than 1 second for a prompt critical LWR,

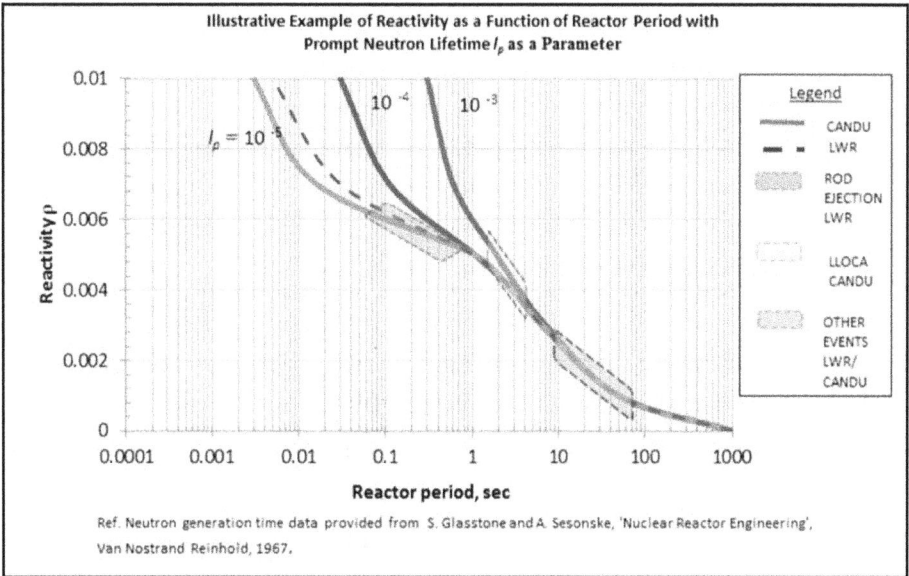

Figure 3 Reactor Period vs. Reactivity for Various Prompt-Neutron Lifetimes

making the rate of power increase (log rate = $1/T$) unacceptable at about 100% P·P·s^{-1}. Because of CANDU's relatively long prompt-neutron lifetime l_p, the CANDU reactor period shows much smaller sensitivity for reactivity insertions > (β - 1 mk); there is no dramatic "cliff-edge" effect around prompt criticality in CANDU, as for other types of reactors.

In safety analysis, the effects of delayed neutrons must be taken into account when analyzing rapid power transients. An accurate analysis requires a sophisticated numerical treatment, such as three-dimensional (3D) space-time neutron-kinetics coupled with multiple-physics feedback. This will not be described here, as it is beyond the scope of this monograph. Interested readers can refer to the books listed in the bibliography at the end of the monograph.

3.5 Reactor Control and Reactivity Control

Most of the time it is desired to maintain the reactor critical so as to keep power steady, at a desired power level. To increase or decrease power, the reactor must be made supercritical or subcritical for some length of time until the desired power is reached, after which criticality must be regained. To shut the reactor down, subcriticality must be achieved and maintained. What must be avoided are situations where the reactor is supercritical for too long, since a divergent power "excursion" may lead to overheating of the fuel, causing melting or other damage.

Two control modes are used in a CANDU plant:

- Reactor leads: the Reactor Regulating System (RRS) controls the plant power to a power requested by the plant operator; and
- Turbine leads: the Unit Power Regulator software (turbine load control) controls the plant's electrical power to a value requested by the electrical grid.

Neutron power has to be controlled locally and globally in order to maintain the power below the prescribed safety limits, to avoid spurious transients and to ensure a rapid reactor shutdown. Neutronic-power control is done via reactivity control. Thus, for effective reactor control, it is important that there be available means of changing and controlling the system reactivity. According to the definition of the multiplication factor, the chain reaction may be adjusted by changing one or more of the following three terms:

- production:
  - adjust the amount of fissile material in the core;
  - adjust the effectiveness of the moderator.
- absorption:
  - use movable absorbers (such as control rods)
  - add or reduce absorbing material in the moderator (soluble poisons);
  - design the fuel with solid, fixed absorbers (burnable poisons) which "burn out" gradually by neutron absorption over the residence time of the fuel in the reactor.
- leakage:
  - change the system dimensions or densities, and/or modify the effectiveness of neutron reflection.

Every reactor must have reactivity devices (control rods and shutdown systems) to control reactivity in both the positive and negative directions, so as to ensure that the fission chain reaction is properly controlled, and that any postulated accidents are mitigated. For a chain reaction to be stable, the net reactivity must be set and maintained at precisely zero. Control of the chain reaction is the primary function of the RRS; effective stoppage of the chain reaction is the function of the shutdown systems (SDS).

In fact, reactor control has several functions:

- Compensating for long-term reactivity changes due to fuel burnup, i.e., variations in the concentrations of heavy nuclei and fission products (this will be discussed in Section 4).
- Compensating for short-term (second-by-second) reactivity and power changes via the RRS (this will be discussed in Section 5). The short-term reactivity and power changes could be caused by xenon effects (to be discussed in Section 7) or local-parameter reactivity feedback (to be discussed in Section 8).

As for safety:

- In the event of an incident, the operator may be able to stop the chain reaction very quickly by inserting a strong neutron absorber. In fact, however, safety does not depend solely on the alertness of the human operator. Every reactor also has fast automatic SDS to be used as soon as the monitoring systems detect a malfunction (this will be discussed in Section 5.2.5 and Section 6).

As will be discussed in the following sections, the control functions can all be performed by a single system, or each function can be performed by a dedicated system. The emergency SDS, however, is always run by a dedicated system.

4. Long-Term Reactivity Change and Control: On-Power Refuelling

4.1 Fuel Irradiation and Fuel Burnup

There are 2 concepts related to the "age" of fuel: irradiation (fluence) and fuel burnup.

The fuel irradiation in a given fuel bundle, denoted ω, is defined as the time integral of the thermal flux in the fuel during its residence time in the core. Another term for irradiation is fluence. Irradiation is also known as the thermal-neutron exposure of the fuel. The units of irradiation are neutrons/cm^2, or more conveniently, neutrons per kilobarn, n/kb. Since the cut-off of the thermal-energy range may be defined differently in different computer codes, the fuel irradiation may vary from computer code to computer code, and caution must therefore be exercised when comparing irradiation values using different codes. In documents, it has been more and more usual to report values of fuel burnup rather than fuel irradiation, as burnup does not suffer from differences in definition between codes.

The fuel burnup in a given fuel bundle is defined as the total fission energy released in the fuel bundle since its entry in the reactor core, divided by the initial mass of heavy element in the bundle. For CANDU, the initial mass of heavy element is the mass of uranium in the bundle. The two most commonly used units for fuel burnup are megawatt-hours per kilogram of uranium, i.e., MWh/kg(U), and megawatt-days per megagram (or tonne) of uranium, i.e., MWd/Mg(U).

The fuel burnup in the bundle as it exits the reactor is called its discharge (or exit) burnup. The average fuel discharge burnup is essentially the inverse of fuel consumption, which is the amount of fuel used to produce a given quantity of fission energy. For a given type of fuel and reactor, the higher the discharge burnup, the lower the refuelling rate or (for LWR) the longer the cycle length, leading to a lower fuel cost. Uranium utilization (the inverse of fuel consumption) is good in CANDU because of its high neutron economy and lack of fuel enrichment.

A typical average fuel discharge burnup attained in the CANDU 6 is 7100-7300 MWd/Mg(U), or 170-175 MWh/kg(U).

The average fuel discharge burnup attained depends on the operational parameters of the core. Any neutron loss or parasitic absorption which reduces the core reactivity will have a negative effect on the attainable fuel burnup. For example, the CANDU 6 has 21 adjuster rods with a reactivity worth of approximately 15 mk. A CANDU reactor which is designed without adjusters or with fewer adjuster rods has a higher excess reactivity and therefore provides higher discharge burnup.

The relationship between reduction in core reactivity and loss of burnup is found to be: 1-mk reduction in the core reactivity leads to a loss of about 120 MWd/Mg(U) in the average fuel discharge burnup.

4.2   Long-Term Reactivity Change with Fuel Burnup

Fuel burnup (or irradiation) has a major effect on the fuel's nuclear properties (and hence on the core reactivity). This is due to the evolution of the fuel's isotopic composition with burnup. The essential phenomena are the depletion of U-235, the conversion of U-238 to Pu-239, fissions in Pu-239 and Pu-241, and the appearance of fission products, as shown in Figure 4. The change of fuel composition versus irradiation (or burnup) is usually calculated in the lattice code by solving the burnup equations, also called the Bateman equations, for each type of fuel bundle.

If we wish to consider the inherent nuclear properties of a given fuel lattice, we can use the concept of the "infinite lattice", defined as an infinite array in 3D of identical lattice cells with the same fuel composition. Note that there is no leakage from an infinite lattice, since it has no boundary. Also, interstitial reactivity devices (see Section 5) are not considered as part of the infinite lattice, since they are present in only some parts of the reactor.

The multiplication factor of the infinite lattice, written as k-infinity or k_∞, is a fundamental property of the specific lattice being considered. It is calculated by a lattice code, based on the nuclear properties of the lattice components. The variation of lattice reactivity with

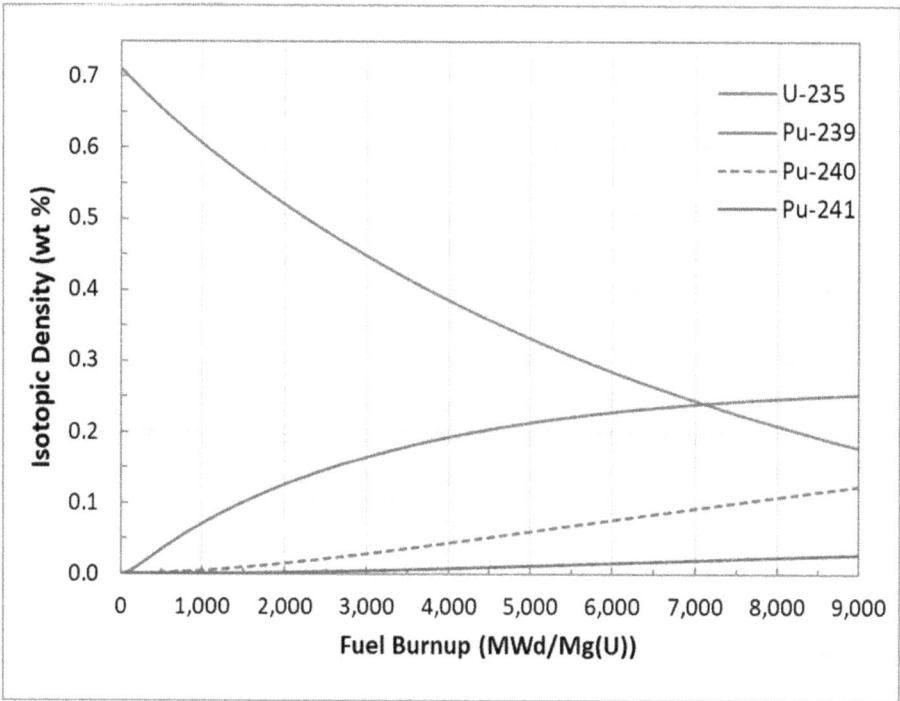

Figure 4 Evolution of Fuel Isotopic Densities with Fuel Burnup for a CANDU-6 Lattice

fuel burnup for the 37-element natural-uranium CANDU lattice is shown in Figure 5.

Figure 6 shows the more important components of the change in reactivity with irradiation (fuel burnup). Initially, as indicated by the combined curve (U-235 + Pu-239), the positive reactivity contribution of Pu-239 overcomes the depletion of U-235 with burnup, so that the net effect is an increase in reactivity (note that the thermal-fission cross section of Pu-239 is ~742 barns compared to ~580 barns for U-235).

Eventually, the net Pu-239 buildup rate slows, because while it is still being produced, it also participates in the chain reaction. The overall effect is that the Pu-239 continues to increase, but the net increase gets smaller and smaller, and this can no longer compensate for the continuing depletion of U-235 burnup, and the curve turns over. In addition,

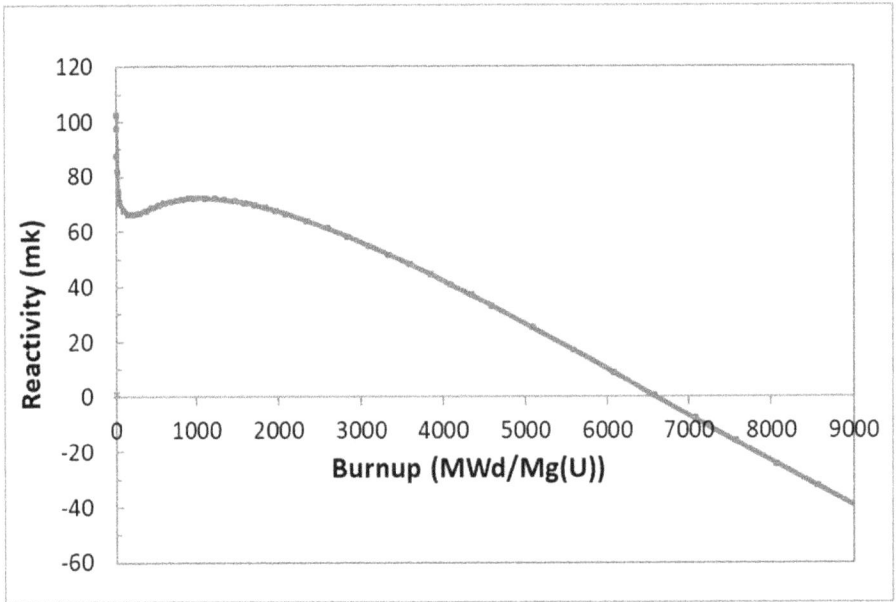

Figure 5 Variation of Lattice Reactivity with Fuel Burnup for the 37-Element Natural-Uranium CANDU Lattice

the steady overall buildup of fission products and of Pu-240 in the fuel makes the decrease in reactivity steeper. The buildup of Pu-241 reduces that rate of reactivity loss somewhat, but is unable to reverse it.

The net effect of all these contributions is shown by the "total" reactivity curve in Figure 6. The peak value in the curve (at an irradiation near 0.5 n/kb) is known as the plutonium peak. Note that this term is somewhat misleading: the peak is not in the concentration of plutonium, but rather it is only a peak in reactivity. A comparison with Figure 5 shows it is essentially the same as the curve for lattice reactivity.

Incidentally, the initial dip in reactivity seen in Figure 5 comes from the accumulation of the saturating fission products Xe-135 and Sm-149, which build to near their equilibrium value over the first week or so of operation (more on this in Section 7). In particular, the very strongly absorbing fission product xenon-135 (not included in Figure

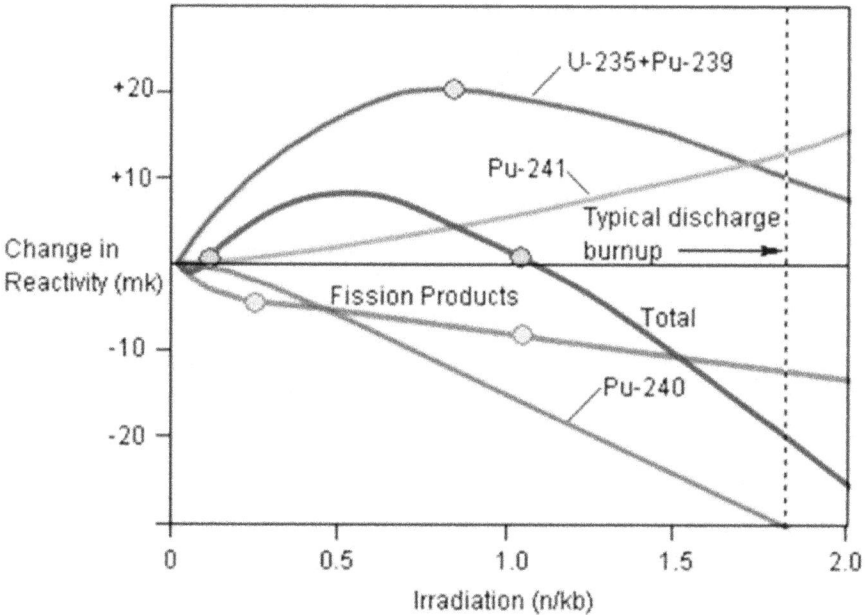

Figure 6 Components of Change of Reactivity with Irradiation (*Source:* Figure 6.5 from [4])

6) builds up in about 40 hours, and this initial buildup of xenon produces the very sharp 28 mk drop in reactivity, as seen in Figure 5.

The equilibrium critical core (with $k_{eff} = 1$) consists of a large number of bundles with different k_∞ values determined by their individual degrees of fuel burnup. The lower k_∞ values of the more highly irradiated bundles are compensated by the higher k_∞ values of the less irradiated ones. In the equilibrium-fuelled core, the net lattice reactivity drops steadily with irradiation. The CANDU-6 reactivity, for example, drops by about 0.4 mk per full-power day. To restore this 0.4 mk of reactivity each day, refuelling must be performed on a daily or close-to-daily basis. In the CANDU 6, some (but not all) of the fuel in 2 or 3 fuel channels is replaced per day, on the average. If the fuelling machines are unavailable for some reason, the reactor can continue to operate for only a limited time. For example, if the average liquid-zone-controller (LZC, described in Section 5.2.1) water level is at 50% full, about

2 mk of excess reactivity is available to counter the 0.4-mk loss per day. This would consequently allow about five days of operation by gradual reduction of the liquid-zone level.

4.3   On-Power Refuelling

As discussed above, CANDU reactors use daily or neart-daily on-power refuelling to maintain long-term criticality and power shape. This keeps the amount of fissile material nearly constant by replacing irradiated fuel bundles with fresh fuel bundles more or less continuously. Two fuelling machines are located at the axial ends of the reactor; they remove a number of used fuel bundles and insert an equal number of fresh fuel bundles. Each fuel channel in the CANDU 6 contains a string of 12 fuel bundles;in some other designs (e.g., Bruce reactors), a channel contains 13 fuel bundles. If the 8-bundle-shift scheme is adopted, the 8 bundles near the outlet end of the channel are discharged, and the 4 or 5 bundles previously nearest the inlet end are shifted nearest to the outlet end. Thus, these 4 or 5 low-power bundles stay in the core for two refuellings of the channel, but the high-power bundles at the centre of the channel leave the core at the next refuelling.

While there are costs associated with on-power refuelling (capital cost and maintenance cost of the fuelling machines), this system has several distinct advantages over batch refuelling, used in other types of reactors:

- "Just-in time" refuelling, helping to maximize fuel utilization;
- No downtime for refuelling;
- Easy removal of failed or defective fuel from the core without reactor shutdown, preserving thereby the purity of the PHTS;
- Better and constant overall spatial flux shape in the core (small local flux variations associated with the fuel burnup and daily refuelling);
- Low excess reactivity, limiting the amount of reactivity increase available in an accident, avoiding the need in batch refuelling for high poison concentrations to hold down reactivity at the beginning of each cycle;

- Low reactivity rates and small ranges (the overall excess reactivity of the core hardly changes from day-to-day and month-to-month)

In addition, CANDU reactors offer extreme flexibility in refuelling schemes:

- The refuelling rate (or frequency) can be different in different regions of the core, and in the limit can in principle vary from channel to channel. By using different refuelling rates in different regions, the long-term radial power distribution can be shaped and controlled.
- The axial refuelling scheme could be 8-bundle shift in one region, and a mixed rate (e.g., 4-bundle and 8-bundle shifts) in another region. And the axial refuelling scheme is not fixed: it can be changed at will, it can be different for different channels, and it does not even need to be the same every time for a given channel.

4.4 Representative Core States During Operating Life of Reactor

From the point of view of fuel management, the operating life of a CANDU reactor can be separated into different periods. The transitional periods are short lasting for 400 to 500 full-power days (FPD) after initial startup, while the "equilibrium core" represents about 95% of the lifetime of the reactor.

Initial fuel load of a new reactor

The initial fuel load of a new CANDU reactor is entirely fresh natural-uranium fuel (that is, fuel with no plutonium or fission products present), except for a small number of depleted-uranium fuel bundles at specific core locations, designed to help flatten the power distribution. Poison in the moderator compensates for the excess reactivity of the fresh fuel in the first few months or so of operation.

Plutonium peak

At about 40-50 FPD of reactor operation, the core reaches its "plutonium peak", at which time the core reactivity is highest, due to the result of the production of plutonium by neutron capture in U-238, and the as-yet relatively small U-235 depletion and fission-product concentration. Following the plutonium peak, the plutonium production can no longer compensate for the accumulation of fission products, and the excess core reactivity decreases.

Onset of refuelling

After 100 to 150 FPD from first criticality, core reactivity drops to a level where a poison shim is no longer required and routine replacement of high-burnup fuel with fresh fuel becomes necessary to maintain core reactivity.

Equilibrium Core

Fuel is replaced on a daily basis (about 16 bundles per day) to add reactivity at a rate equal to the rate of loss from burnup. During the transitional period which follows, the reactor gradually approaches the "equilibrium" state, approximately 400 to 500 FPD after initial startup. The overall refuelling rate, the in-core average burnup, and the burnup of the discharged fuel have become essentially steady with time. The global flux and power distributions can be considered as having attained an equilibrium, "time-average", shape, about which the refuelling of individual channels leads to local "refuelling ripples".

4.5 Reference Time-Average Flux/Power Shapes and Refuelling Ripples

As described above, the long-term operation of a CANDU core is characterized by an equilibrium-core spatial flux distribution which is overall constant in time, with localized perturbations due to refuelling of channels. The equilibrium-core flux shape as designed by analysis is called the "time-average" model, whose output becomes the reference target shape for the core power. The "refuelling ripple" defined for each channel at any given time is the ratio of the channel's instantaneous power to its value in the reference channel-power distribution. These ripples are due to the various instantaneous values of fuel burnup in the different channels, which are the result at any given instant of the specific sequence in which channels have been refuelled. Carrying out a CANDU-6 time-average calculation is essential at the design stage. Table 3 shows typical results.

4.6   Channel-Power Peaking Factor (CPPF)

Because many safety analyses are normally carried out based on a time-average power distribution, it is very important to quantify how much higher the instantaneous power distribution peaks above the time-average power distribution. This is given by the Channel-Power Peaking Factor (CPPF), defined as the highest ripple over all channels in the "CPPF Region", which typically excludes low-power rings of channels near the periphery of the core. The CPPF varies from day to day. A

Table 3   Typical Results from a CANDU-6 Time-Average Calculation (*Source*: Table 2 from [3])

CANDU 6	Time-Average Model
Total Reactor Power (MW)	2061.4
Total Fission Power (MW)	2158.5
Average Channel Power (kW)	5425
Average Bundle Power (kW)	452
Uranium Mass per Bundle (kg)	19.1
Maximum Channel Power (kW)	6604.2 (N-6)
Maximum Bundle Power (kW)	805.2 (O-5, Bundle 7)
k_{eff}	1.00250
Average Exit Irradiation (n/kb)	1.706
Average Exit Burnup	171.3 MWh/kg(U); 7139 MWd/Mg(U)
Avg. Channel Dwell Time (FPD)	192
Feed Rate	1.98 Channels/FPD; 15.83 Bundles/FPD
Reactivity Decay Rate (mk/FPD)	-0.385
Average Zone Fill (%)	50.0
Coolant Purity (atom %)	99.0
Moderator Purity (atom %)	99.833
Number of Burnup Zones	5

typical value of the CPPF averaged over time is ≤1.10, depending on the axial refuelling scheme used. The greater the number of bundles replaced per channel at each refuelling, the greater the reactivity increment, and therefore the greater the refuelling ripple, and the higher the CPPF. The exact value of the CPPF must be monitored because it is used to calibrate the in-core detectors in the Regional/Neutron Overpower Protection (ROP/NOP) systems, which will be discussed in detail in Section 6.

4.7   Selecting Channels for Refuelling

One of the main functions of the fuelling engineer (or station physicist) is to establish a list of channels to be refuelled during the following period (few days) of operation. Normally, channel selection will begin with eliminating channels which are poor candidates for refuelling. With experience, a fuelling engineer will develop a personal set of rules for eliminating channels. Once channels inappropriate for refuelling have been eliminated, possible lists can start to be developed from the remaining channels. The fuelling engineer will usually have to draw up a list from many options available.

A computer code that simulates the core (such as RFSP or SORO) helps select channels for refuelling. The fuelling engineer reviews the output of the computer code and then selects channels for refuelling based on safe operation, reactivity requirements, flux shape control, fuel cost, and efficient use of the refuelling equipment. A good way of being confident about a channel selection is to perform a pre-simulation of the core following the refuellings. This pre-simulation (especially if it invokes bulk- and spatial-control modelling) will show whether the various power, burnup, and liquid-zone-fill criteria are likely to be satisfied, or whether the channel selection should be changed.

4.8   Compliance with License Limits on Channel and Bundle Powers

Upper power limits for fuel channels and fuel bundles must be met during the operation of a nuclear reactor (license conditions). For a

CANDU 6 reactor using the 37-element fuel bundle, the maximum channel power is 7.3 MW and the maximum bundle power is 935 kW. The channel power limit is designed to avoid dryout in most accidents while the bundle power limit is designed to minimize fission-product releases to the environment in case of a fuel element rupture. Channel overpower can result in fuel-sheath dryout, which will decrease heat transfer from fuel to coolant and increase fuel temperature and possibly cause fuel damage. Bundle overpower can lead to centreline melting even if there is adequate cooling, which will likely cause sheath failures.

Fuel dryout must be avoided in all channels. Accordingly, the regulating and safety systems must provide full protection against fuel dryout and require thorough logic and redundancy. As will be shown in Section 6, an ROP/NOP trip ensures the integrity of the fuel channel by reducing power before there is risk of fuel dryout. However, the ROP/NOP systems do not prevent the violation of license limits on bundle and channel powers.

Enforcement of the license limits on channel and bundle powers requires the joint efforts of the fuelling engineer, the operator, and the authorized staff. The fuelling engineer routinely checks the slow evolution of steady-state bundle and channel powers via core-follow calculations with the computer code (such as RFSP or SORO). Neutron power has to be controlled locally and globally routinely via the RRS and on-power refuelling in order to maintain the power below the prescribed safety limits, to avoid spurious transients and to ensure a rapid reactor shutdown. This means that routine monitoring of device positions and liquid-zone levels is part of the enforcement process for license limits on bundle and channel powers. Conversely, if the flux shape is not standard (for example, there is an unusual tilt), the operator has little choice but to assume that the channel-power limit (or bundle-power limit) may be exceeded, unless analysis of the particular configuration demonstrates otherwise.

4.9   Flux/Power Flattening

The total power output of the reactor is proportional to the average flux, so it is advantageous that this be as high as possible. Operation of a reactor at full power with a "peaked" power shape, as for instance with a power shape found in the analysis of homogeneous reactors, may result in some channel powers being higher than the maximum value allowed by the operating license. The total reactor power can be maintained by redistributing power from high-power channels to lower-power channels, i.e., by ensuring flux and power flattening. This flattening is achieved in a CANDU reactor in the radial and axial directions in the following ways:

1. Addition of a reflector (for neutron-leakage reduction and for radial flattening);
2. Use of adjuster rods to absorb neutrons in the core centre to provide axial and radial flattening (note that another function of the adjuster rods is to provide excess reactivity when needed);
3. Use of bi-directional refuelling (for axial flattening);
4. Use of differential discharge burnup, i.e., use of different refuelling rates in different radially concentric regions (for radial flattening).

The reactor-design process implements the first two methods; the refuelling strategy achieves the last two. The ultimate justification for flux flattening is an economic one. With the various flux-flattening strategies in the CANDU technology, the net result is approximately doubling the heat output of the reactor without breaking compliance with license limits on channel and bundle powers.

5. Short-Term Reactivity & Power Control: Reactor Regulating System (RRS)

5.1 RRS Main Components and Functions

For CANDU reactors, the control of the long-term reactivity and of the power is carried out by on-power refuelling, while the control of the short-term reactivity and of power is done by the RRS. The RRS is part of the overall plant-control system that maintains the reactor power at a specified level, or, when required, manoeuvres the reactor power between specified setpoints. The reactor power setpoint can be entered by the operator (in the reactor-leading mode) or it can be calculated automatically by the Steam Generator (SG) pressure-control program (in the turbine-leading mode). The RRS consists of the following main components:

- Reactivity-control devices such as LZCs, mechanical control absorbers (MCAs), adjuster rods, and moderator poison;
- Reactor-regulating computer programs such as power measurement & calibration, demand power routine, reactivity-device control, setback routine, stepback routine, flux-mapping or Fully INstrumented CHannel (FINCH) system;
- Control instrumentation (input sensors) such as ion chambers, process-control instrumentation, in-core flux detectors or FINCH detectors; and
- Hardware interlocks, plus a number of display devices.

The RRS combines the reactor's neutron-flux and thermal-power measurements, reactivity-control devices, and its set of computer programs to perform three main functions:

- Monitor and control the total reactor power to satisfy station-load demands;
- Monitor and control the reactor flux shape; and
- Monitor reactor power and reduce it at an appropriate rate if any parameter is outside specific limits.

Except for extremely low power levels (below 10^{-7} of full power), RRS action is fully automatic, by digital routines which process the inputs and drive the appropriate reactivity-control devices. A general block diagram of the RRS is shown in Figure 7. The Power Measurement and Calibration Routine uses measurements from a variety of input sensors to arrive at calibrated estimates of bulk and zonal reactor powers. The Demand Power routine computes the desired reactor power setpoint and compares it with the measured bulk power to generate a bulk-power error signal that is used to operate the reactivity devices.

The primary control devices are the 14 LZCs. The levels of light water in the LZCs (also called "liquid-zone levels" in the rest of the monograph) are varied in unison (for bulk-power control) or differentially (for spatial-power control). If the reactivity required to maintain the reactor power at its specified setpoint exceeds the capability of the LZCs, the RRS calls on the other reactivity devices. Adjuster rods are withdrawn from the core for positive reactivity shim (note

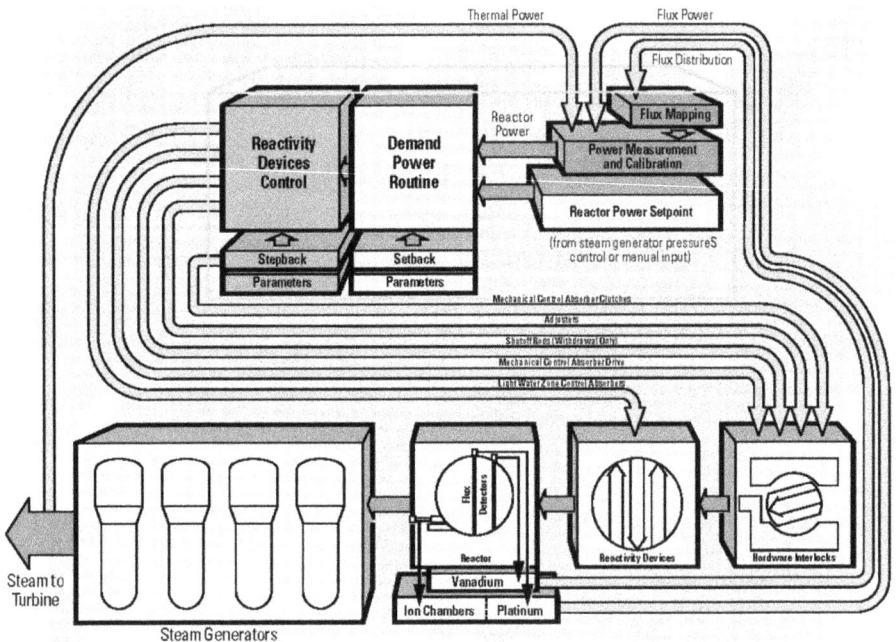

Figure 7   The Reactor Regulating System (RRS) for CANDU 6 (*Source*: [7])

that this function is inhibited if the reactor power setpoint is above 97%). Negative reactivity is provided by inserting MCAs or by adding poison to the moderator. The movement of these devices is dictated by the average liquid-zone levels and the effective power error. The approximate reactivity worths of these devices, and the highest average rates at which the reactivity can be added or removed, are shown in Table 4.

In addition to controlling the reactor power, the RRS monitors a number of important plant variables and reduces reactor power, if any of these variables is out of limits. This power reduction may be fast (stepback), or slow (setback), depending on the possible consequences of the variable lying outside its normal operating range. Stepbacks are accomplished by dropping the MCAs, while in setbacks the MCAs are driven in response to a reduced power setpoint.

The reliability of the RRS is of the utmost importance. The frequency of serious process failures associated with the RRS, i.e., loss of regulation (LOR), must be kept to a minimum (< 1 in 100 years). The design intent is that there be no LOR during an earthquake of intensity less than or equal to the design-basis earthquake.

Table 4 CANDU-6 Reactivity-Device Worth (*Source*: Based in part on Table 4.1 from [5] and Table 5.1 from [6])

Function	Reactivity Devices	Total Reactivity Worth	Maximum Insertion Rate
Control	14 LZCs	-~7 mk	±0.14 mk/s
Control	21 Adjusters	-~15 mk	±0.10 mk/s
Control	4 MCAs	-~10 mk	~140 s (driving) ~3 s (dropping)
Control	Moderator Boron	-8 mk/ppm	variable
Control	Moderator Gd	-28 mk/ppm	variable
Safety	SDS1	-50 mk to -80 mk	~1 s
Safety	SDS2	-300 mk	~2.3 s

5.2 Reactivity-Control Devices

Reactivity devices are located in the moderator, a benign environment (see Figure 8), so no rod-ejection event can occur in a CANDU reactor. Also, reactivity devices are not subject to damage, other than in-core pressure-boundary breaks. The total reactivity worth invested in the RRS is low (~15 mk). The rod worth of single rods, and the total reactivity change due to any malfunction, are limited.

The reactivity devices used for reactor control, are described in detail below. In addition, it should be noted that withdrawal of the safety-system shutoff rods (SORs) is carried out under computer control by the RRS. In this section we will describe the functions of the different reactivity-control devices.

5.2.1 Liquid Zone Controllers

The LZCs are the primary reactivity-control mechanism for fine control of reactivity. Six tubes within the reactor core (see Figure 9) contain 14 compartments into which light water is introduced. The water acts

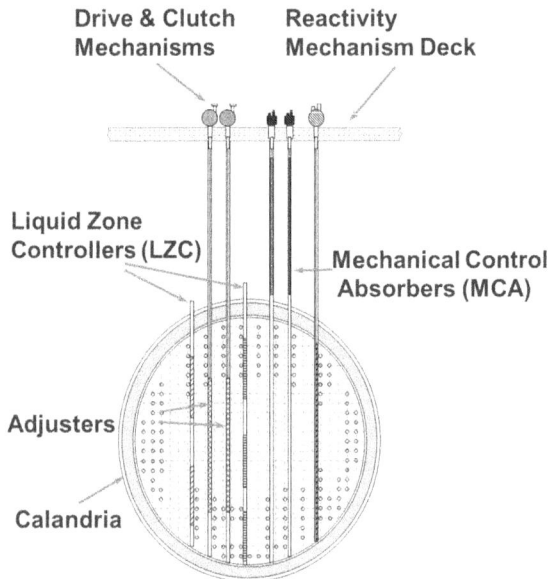

Figure 8   End-Elevation View of Reactivity-Device Locations

as a neutron absorber, and the water level or fill in each compartment is controlled by manipulating the compartment's inlet valve. The water is forced out of the compartment at a constant rate by the helium-cover-gas pressure. The difference in reactivity between the state with all compartments empty and all compartments full is about 7 mk. The rate of change of reactivity when all 14 absorbers are filling or draining in unison at their maximum rate is about ±0.14 mk/s. The LZCs are designed to accomplish the following two functions:

- Bulk Control: All water fills are varied together in the same direction every 0.5 s, to keep the reactor critical for steady operation, or to provide small positive or negative reactivity to increase or decrease power in a controlled manner.
- Spatial Control: Water fills are varied differentially every 2 seconds in the individual compartments, to shape the 3-D power distribution towards a desired reference shape.

The LZC system is preferentially operated at or near 50% average water fill to provide the capability for reactivity control of about ±3 mk, since this is sufficient to compensate for routine reactivity perturbations due to refuelling.

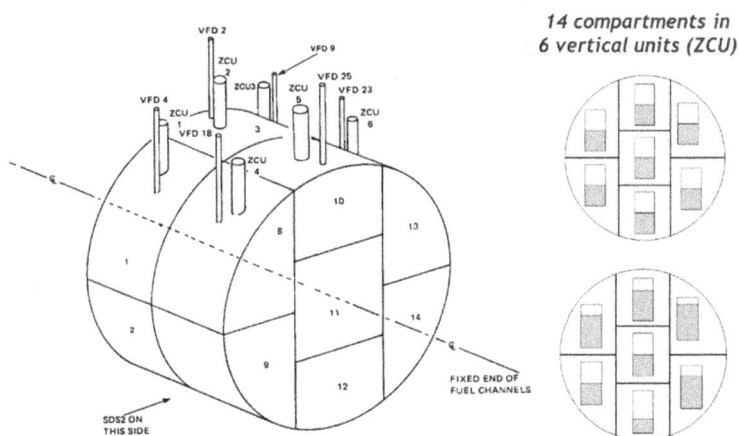

Figure 9   Liquid-Zone-Control Units

5.2.2 Adjuster Rods

The CANDU 6 reactor has 21 adjuster rods, made of a neutron-absorbing material (stainless steel, or cobalt[1]) which are normally fully inserted into the central regions of the reactor core to reduce and redistribute the flux peak. Being symmetrically distributed in banks around the core center, adjuster rods affect both radial and axial flux shapes. The use of adjuster rods enables a higher total power output for the same maximum flux and local power. As adjuster rods are normally inserted in the reactor at full power, they represent a negative reactivity contribution. To overcome this the CANDU plant must increase the fuel replacement frequency by approximately 10 %.

When called upon for reactivity addition, adjuster rods are withdrawn successively in banks. Instances when this is done are, typically, to add positive reactivity to override a xenon transient or to allow extended power production without refuelling when the fuelling machine is unavailable for a number of days.

If refuelling were to stop, core reactivity would continuously decrease (~0.4 mk/FPD in the CANDU 6). When this happens, the operator would ensure any moderator poison is removed and the RRS would of course attempt to maintain criticality. Specifically, lowering the liquid-zone levels to their lower limit would give 2-2.5 mk of positive reactivity, or ~5-6 extra days of operation. Then withdrawal of some or all adjuster rods, bank by bank, would provide positive reactivity shim (hence it is called shim operation) to permit operation to continue for several weeks.

In situations when some or all adjuster rods are withdrawn to provide positive reactivity (shim operation, startup), the flux shape loses the flux flattening provided by the adjuster rods, and derating is required, in order to remain in compliance with licensed maximum channel-power and bundle-power values. The amount of derating increases with the

---

[1] It is worth mentioning that many CANDU utilities have used cobalt-59 adjuster rods to harvest cobalt-60 (after few years' activation in the reactor core) for industrial and medical use.

number of adjuster rods withdrawn. Conversely, in situations when the liquid-zone levels increase and reach their higher limit, additional negative reactivity is needed, and if some adjuster banks are out of core, they are re-inserted, again one bank at a time.

The reactivity change in the CANDU 6 when all adjuster rods go from fully inserted to withdrawn is approximately 15 mk. The average reactivity rate when two banks of adjuster rods are driven at full speed is approximately ±0.08 mk/s.

5.2.3 Mechanical Control Absorbers

There are 4 MCAs, mechanically similar to SORs, but controlled by the RRS. They can be driven in or out at variable speed, or they can be dropped under gravity by releasing a clutch. The MCAs absorbers are normally out of the core, and are driven in, one bank of 2 rods at a time, to add negative reactivity when needed (i.e., on very high liquid-zone levels), or are dropped to provide a fast reactor power reduction (stepback). Unlike adjuster rods, MCAs are driven in and kept in only as long as the condition of very high liquid-zone levels persists.

The MCAs have sufficient reactivity worth to compensate for the initial-core fuel-temperature coefficient of reactivity on shutdown from full power. The MCAs also prevent the reactor from going critical on SOR withdrawal. This is to enable the shutdown system one (SDS1) to be repoised before criticality.

The reactivity change when all MCAs go from fully inserted to withdrawn is approximately 10 mk. The average reactivity rate when driving the rods at full speed is ±0.07 mk/s. When dropped, the rods take approximately 3 seconds to insert fully. By re-energizing the clutch while the rods are dropping, a partial insertion to any intermediate position can be achieved.

5.2.4 Poison Addition and Removal

A reactivity balance can be maintained by the addition of soluble poison to the moderator. Boron is used to compensate for high reactivity due to an excess of fresh fuel, as it does not burn out rapidly. Gadolinium is added when the xenon load is significantly less than at equilibrium (as happens after prolonged shutdowns).

An ion-exchange system removes the poison from the moderator. Initiating the addition or removal of poison is normally done by the operator. The RRS does, however, have the capability of adding gadolinium, if needed, to compensate for gross errors in the reactivity balance.

5.2.5 Shutdown System One (SDS1) and Shutdown System Two (SDS2)

CANDU reactors have two shutdown systems, SDS1 and SDS2. The SORs associated with SDS1 and the liquid-injection nozzles associated with SDS2 basically control reactivity in one direction only. SDS1 and SDS2 are designed to be both functionally different and physically separate from each other (see Figure 10).

The SDS1 SORs utilizing cadmium absorber elements are designed to quickly shut down the reactor under normal and emergency

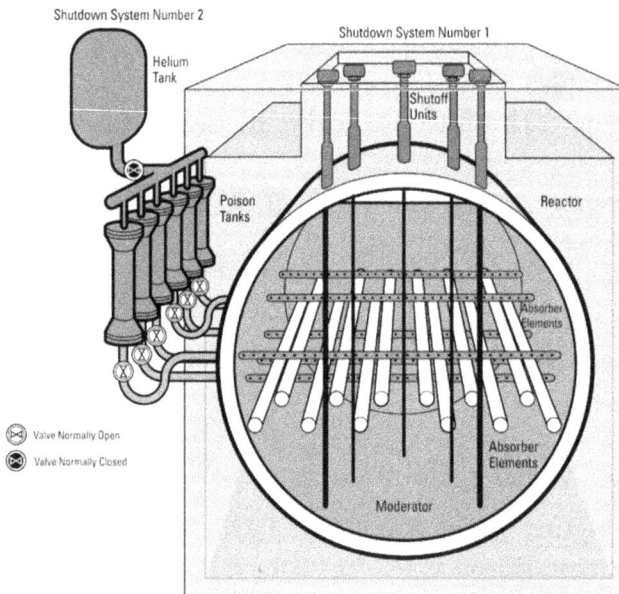

Figure 10   General Layout of SDS1 and SDS2 (*Source:* [7])

conditions. The rods, being diverse in design principle and orientation from SDS2, hang above the core, suspended from the reactivity-mechanism deck. 28 SORs are used in CANDU-6 reactors. They fall into the reactor under gravity, inside a perforated zirconium alloy guide tube within the core, in response to a shutdown signal. SOR withdrawal is normally done under control of the RRS. Thus the SORs can be considered as providing reactivity control as part of the RRS, only when being operated this way.

The SDS2 operates by injection under high pressure of the neutron-absorber gadolinium (as gadolinium nitrate in high concentration in heavy water), directly into the moderator. Figure 10 show a schematic of the liquid-injection system and the location of the injection nozzles for CANDU- 6 reactors. The Zircaloy-2 nozzles penetrate the calandria horizontally and at right angles to the fuel channels.

As previously discussed, it is a safety concern if a large positive reactivity is allowed to occur, or if a positive reactivity is maintained for a long time. Automatic reactor shutdown occurs when instruments detect a high rate of power increase. Then reactor stepback drops MCAs into the core, a trip on SDS1 drops SORs into the core, or a trip on SDS2 injects gadolinium poison at high pressure into the moderator. These systems trigger at rates corresponding to the following reactor periods T:

- $T = 12.5$ s (log rate = 8% $P \cdot P \cdot s^{-1}$) for stepback,
- $T = 10$ s (log rate = 10% $P \cdot P \cdot s^{-1}$) for SDS1, and
- $T = 6.7$ s (log rate = 15% $P \cdot P \cdot s^{-1}$) for SDS2.

Note that SDS can also be actuated if the power level measured by in-core detectors reaches a high value (see Section 6).

5.3 Reactor Regulating Computer Programs

Reactor-regulating computer programs are executed through dual reliable and redundant digital-control computers (DCCs), which provides defenses against LOR. This section describes the main features of the reactor-regulating computer programs.

5.3.1 Power Measurement and Calibration

The RRS requires power measurements to control zone power and total reactor power. A fast, approximate estimate of reactor power is obtained by taking either the median ion-chamber signal (at powers below 5% FP), or the average of self-powered in-core flux detectors (above 15% FP), or a mixture of the two (between 5% and 15% power). These measurements generate short-term error signals to drive the LZCs and stabilize the core. Over a long time span, these signals are slowly calibrated to agree with more accurate estimates of reactor power and zone power.

5.3.2 Demand-Power Routine

The demand-power routine serves 3 functions: it determines the mode of plant operation (turbine leading, reactor leading, or setback); it calculates the reactor setpoint, and the effective power error that is used as the driving signal for the reactivity control devices; and it adds poison to the moderator if required. The effective power error is calculated as a weighted sum of the difference between the set and measured flux powers, and the difference between the set and measured rates.

5.3.3 Setback and Stepback Routines

The setback routine monitors a number of plant variables and reduces reactor power in a ramp fashion if any variable exceeds predefined operating limits. The rate at which reactor power is reduced and the power level at which the setback ends, are chosen to suit each variable. The setback clears either when the end point is reached or when the condition clears. Setbacks override any operator requests for power setpoint. Also, if the plant is running in reactor-leading mode, the setback changes the operating mode to turbine-leading mode.

The stepback routine checks the values of core variables, in particular the neutron flux and the log rate. If these are sufficiently high, a fast power reduction is necessary, and the stepback routine then opens the clutch contacts of all four MCAs. The MCAs are dropped either partly or fully into the core, effecting the rapid power drop. Most stepbacks lead to complete shutdowns, but a few conditions require only a partial fast power reduction. Stepback on high neutron flux or high log rate serves to prevent a LOR.

5.3.4 CANDU-6 Flux-Mapping Routine

The CANDU-6 flux-mapping routine collects readings from 102 vanadium flux detectors distributed throughout the CANDU-6 core. It computes a best fit of these readings by using a linear combination of basis functions known as flux modes (harmonics modes and device modes) expected for the given core configuration. The process is called mapping the flux distribution. Flux mapping is very fast, and it provides an accurate estimate of the 3D power (actually neutron-flux) distribution, from which the average zone flux in each of the 14 zones in the CANDU 6 reactor can be computed. These estimates are available once every 2 minutes (the flux-mapping sampling interval) and lag the neutron flux on account of the 5-minute time constant of the vanadium-detector response. The calibration of 14 pairs of platinum-clad zonal flux detectors is done by matching the appropriately filtered (t = 5 min) detector reading to the average zone-flux estimate generated by flux mapping. Note that Ontario's CANDU reactors use the FINCH system instead of flux mapping.

5.4   Control Instrumentation (Input Sensors)

Neutron flux in a CANDU reactor spans a range of many decades. The flux in a fresh core prior to first criticality is approximately 10^{14} times lower than the flux in an equilibrium core at full power. Because of this large range, different types of instruments are required, depending on the power level. These measurements are required both for control of reactor power and for reactor protection.

5.4.1 Startup Instrumentation

At very low power levels (from $\sim 10^{-14}$ to 10^{-6} of full power), power is measured by supplementary "startup instrumentation" (BF3 gas counters, or He-3 counters at some stations) located temporarily in-core in the calandria viewing port, and also out-of-core in the SDS2 ion-chamber housings. The in-core counters are needed only for the initial reactor startup (from $\sim 10^{-14}$ to 10^{-10} of full power); the out-of-core counters may be needed for restarts following extended reactor shutdowns as well as for the initial startup. High accuracy of calibration is not required since it is sufficient to obtain an indication approximately proportional to the actual reactor power. However it is important that an output be provided which is always on-scale and which responds promptly to changes in the power level. The instrumentation must trip SDS1 if reactor power or rate-of-change of power exceed preset limits, or if the startup instrumentation or its power source should fail.

5.4.2 Ion Chambers

At power levels of about 10^{-7} of full power, the permanently installed out-of-core ion chambers start coming on scale. The ion chambers span a range of operation from 10^{-7} to about 150% FP, but they are used only at relatively low reactor powers, below 10% FP, because they are not very accurate and give no information on flux tilts. The ion chambers are sensitive to both neutron and gamma rays. The gamma rays originating from the reactor core produce an unwanted signal, leading to significant background signal noise when the neutron flux is small, such as during a reactor startup at 10^{-6} FP. At full power, the ion-chamber current contributed from gamma rays is about 3 decades below that contributed from neutrons.

The RRS ion-chamber system is a triplicated system consisting of 3 independent logical channels A, B and C. Each channel provides two sets of outputs to dual redundant DCCs and a third set to three chart recorders used to record all 3 signals continuously. The output current from each ion chamber goes to an amplifier, which produces log neutron power, linear neutron power, and log-rate signals. The log-rate signal is a direct trip parameter. The log and linear-power signals are used as conditioning signals for other trip parameters.

5.4.3 In-Core Flux Detectors

The in-core flux detectors (ICFD), also known as self-powered neutron detectors (SPND), are the primary in-core sensors used in CANDU reactors during normal operation at high power levels. The detector is self-powered because it does not require an applied bias voltage to separate and collect an ionization charge to derive a signal.

The primary considerations in the choice of the ICFD emitter types are the speed of response, the kind of sensitivity (to neutrons or gamma rays), and the tolerable burnout rate. In CANDU reactors, the high neutron flux has severely restricted options to emitter types with low burnout rates such as Vanadium, Platinum and Inconel. Currently, CANDU reactors use only 3 types of emitters: vanadium, platinum-clad Inconel, and pure Inconel.

CANDU reactors use a large number of ICFDs for flux measurement, reactor control and protection. For example, a CANDU-6 reactor has 26 Vertical Flux Detector Assemblies (VFDA) and 7 Horizontal Flux Detector Assemblies (HFDA). The 26 VFDAs house 28 platinum-clad Inconel detectors (2 in each zone) for the RRS spatial-power control, 102 vanadium detectors for the RRS flux mapping, and 34 platinum-clad Inconel detectors for the SDS1 Regional Overpower Protection (ROP) system. The 7 HFDAs house 24 platinum-clad Inconel detectors for the SDS2 ROP system.

Vanadium detectors are not suitable for direct use in safety or control systems because they have a dominant time constant of 325 seconds. In response to a step increase in power the vanadium detector signal will take about 25 minutes to indicate 99% of the power increase. However, they are suitable for neutron-flux mapping in CANDU reactors, because a vanadium detector is essentially only neutron sensitive. The vanadium-flux-detector measurements are periodically calibrated when the reactor is at a power level that is accurately known and the flux shape is undistorted. This is done to offset a deterioration of detector sensitivity with irradiation of about 4% per year.

The Inconel detector (used in Ontario's CANDU units) is almost fully neutron sensitive and responds promptly to changes in neutron flux, except for a small (~5%) negative delayed component produced by the manganese impurity in Inconel. The platinum-clad Inconel detector (used in CANDU 6) has a mixed response with ~70% of the signal produced by neutron absorption and ~30% by gamma radiation (one-third of which is delayed gamma rays). The dynamic response of the Inconel detector is ~105% prompt, and the platinum-clad Inconel detector is ~89% prompt. For the less-than-prompt response of the flux detector signals, dynamic compensation is provided to ensure timely trips for fast power increases.

The prompt fraction, delay fraction and delay constants are functions of the detector materials (changing over time due to neutron radiation), the geometry, and also the gamma and neutron spectra. Periodic monitoring of the ICFD performance, including their dynamic response, is recommended for all operating CANDU reactors.

6. Regional/Neutron Overpower Protection (ROP/NOP)

6.1 Description of ROP/NOP Systems

There are independent, separate, and diverse ROP/NOP systems for the two SDS. Each ROP/NOP system consists of a number of flux detectors (see Section 5.4.3) which provide prompt measurements of neutron flux throughout the core. The detectors are mounted inside assemblies that penetrate the core, perpendicular to the fuel channels, vertically or horizontally. The system for SDS1 uses vertical detectors, the one for SDS2 uses horizontal detectors. Detectors are judiciously distributed to monitor the neutron flux throughout the core. In the CANDU 6, a total of 58 ROP detectors are used: 34 for SDS1 and 24 for SDS2. The number and location of detectors in the core are selected in an analysis whose objective is to ensure that as small a number of detectors as possible protect the reactor by tripping a SDS when local high powers threaten reactor safety from any flux shape that could arise in the operating reactor, while at the same time providing adequate margin-to-trip (MTT) to avoid possible restrictions on reactor operating power.

Detectors are grouped into 3 separate logic (or safety) channels in each SDS, 4-18 detectors per logic channel. These channels are labelled D, E, and F for SDS-1 and G, H, and J for SDS-2, as illustrated in Figure 11. A logic channel trips if the reading of any detector in that channel reaches a predetermined setpoint. A shutdown system is actuated whenever 2 out of the 3 corresponding logic channels are tripped (this is called 2/3 trip logic or triplication). The triplication not only provides adequate protection, but also reduces the chance of a spurious trip, and allows the system to be tested on-line. It assures an extremely high reliability of shutdown-system actuation under accident conditions.

6.2 Design Basis of ROP/NOP Systems

The ROP/NOP systems provide primary or backup trip coverage for a variety of process failures that may cause excessively high power.

SHUTDOWN SYSTEM NO. 1 (VERTICAL)

TRIP CHANNELS:

CHANNEL D

CHANNEL E

CHANNEL F

CALANDRIA

FUEL CHANNELS

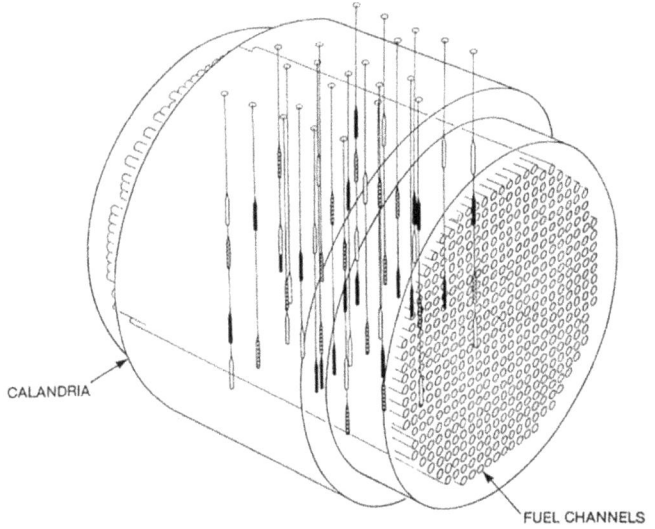

SHUTDOWN SYSTEM NO. 2 (HORIZONTAL)

TRIP CHANNELS

CHANNEL G

CHANNEL H

CHANNEL J

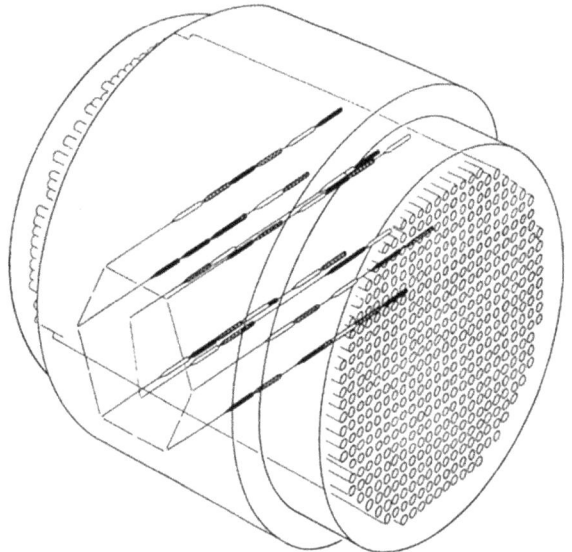

Figure 11 Typical ROP/NOP Detector Layouts in CANDU 6 (*Source*: Figure 3 from [8])

The systems cater particularly to power increases that are too slow to produce a log-rate trip. The design function of the ROP/NOP system is to initiate a reactor shutdown whenever the neutron flux reaches an unacceptably high level anywhere in the reactor core. Such a condition can occur in any postulated design-basis event, from the very unlikely large LOCA to relatively more likely LOR events involving a loss of control of the bulk power and/or of the spatial power distribution in the reactor. An LOR could occur because of failure of the controller (computer or computer program), operator error, failure of the reactivity-control devices, or failure of the power-measurement devices. By design, the probability of an LOR should be less than 1 in 100 years. A major defense against LOR is the stepback function, which is independent of the normal power-control program The stepback routine uses simple, fail-safe devices (the MCAs, which it drops under gravity), and simple, somewhat independent power measurements. Note that the RRS does not actuate the shutdown systems, and it is not credited with the shutdown function in design-basis accidents (DBAs).

During an LOR, if the local power increase in a fuel element is sufficiently large then an unstable dry patch on a fuel sheath may develop. This condition is commonly referred to as dryout. Although the onset of fuel-sheath dryout does not necessarily lead to failures of the fuel or fuel channel, elevated fuel temperatures can result in fuel-element deformations and, in the extreme, in fuel-centreline melting and eventually pressure-tube failure. Therefore, timely actuation of the SDS is critical.

The SDS would be actuated once the ROP/NOP detector signal reaches a pre-established value called trip setpoint (TSP). The TSP of the ROP/NOP detectors is determined by an extensive analysis so that the SDS will be actuated in time to prevent Onset of Intermittent fuel-sheath Dryout (OID). The analysis methodology is referred to as the ROP/NOP analysis methodology, and the DBA scenario for calculation of the TSP is a postulated slow LOR event.

ROP/NOP systems provide protection against slow LORs when the reactor is operating in a wide range of flux shapes, covering normal operating conditions and device configurations, and abnormal device configurations (the core may continue to operate with a distorted power profile). A slow LOR event could start from any of these flux shapes. Since the ROP/NOP trip is generally the only trip parameter available for slow LORs, a very large number of flux shapes, typically ~1000, are examined to challenge the ROP/NOP systems and ensure robustness. The design-basis set of flux shapes is a comprehensive set of flux shapes representing all analyzed upsets.

6.3   ROP/NOP TSP Calculation

The ROP/NOP calculation procedure is summarized in Figure 12. At a generic level, the ROP/NOP computational framework is a ROP/NOP TSP analysis model which includes several computer codes (for reactor-physics and thermalhydraulics) and associated models account for PHTS aging and other information needed to apply the computational framework to a postulated slow LOR event, including procedures for treating the input and output information and related uncertainties, and various approximations needed in the computational procedure.

In the ROP/NOP analysis, the reactor-physics code (such as RFSP or SORO) is used to calculate the ROP/NOP reference channel-power distribution, the instantaneous channel-power distribution, channel overpowers, ROP/NOP detector readings, power-shape parameters (i.e., CPPF, ripples, and ripple correction for xenon-free effects), etc. The thermalhydraulics code (such as NUCIRC or TUF) is used to determine for each channel the power which will lead to OID, i.e., the channel's Critical Channel Power (CCP), and the changes in header boundary conditions that may occur due to a change in power level prior to an ROP/NOP trip, etc. The results from these calculations are subsequently used in the ROP/NOP analysis code to determine the TSP required to ensure that the acceptance criteria are met for the slow LOR.

Another requirement for an ROP/NOP TSP analysis model is to ensure that the MTT is less than the margin-to-dryout (MTD) with a sufficiently high probability. As a result, if the detector response is

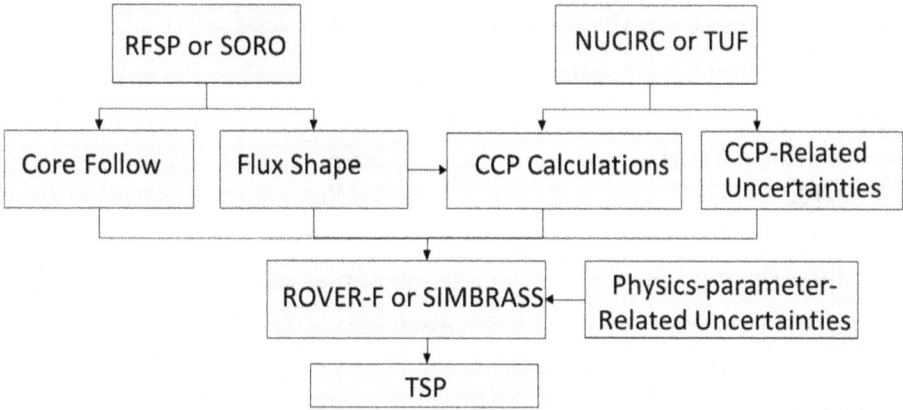

Figure 12 Simplified ROP/NOP Computational Framework (*Source*: Figure 1 from [9])

below the TSP, then all channel powers are below their CCP. Therefore, we have the following conceptual inequality for each ROP/NOP system:

$$MTT = \frac{TSP}{detector\ reading} \leq \frac{CCP}{channel\ power} = MTD$$

The value of TSP that just achieves the inequality protects the idealized reference core, without considering refuelling ripples due to fuel burnup and the sequence of refuellings. It is important to derive the true TSP which corresponds to the actual conditions at the OID. Such conditions, however, cannot be predicted with certainty because it is of course not known when the slow LOR will take place. Therefore, the TSP is actually a stochastic variable defined in terms of other stochastic variables. Hence adjustments are needed to the MTT and MTD in the above inequality to account for uncertainties in the refuelling ripples and in other specific parameters. Specifically, the channel power will increase or decrease for each channel by its ripple, the MTT will decrease by a factor of CPPF for every detector, on account of the calibration process, and all of the inputs in the above inequality have uncertainties related to them. These items are discussed further below.

6.4 Superposition Principle and Detector Calibration

The superposition principle is used to separate the effect of the refuelling ripple from the flux-shape variations (initial ROP/NOP flux shapes), by assuming that the refuelling ripple remains unchanged for different ROP/NOP flux shapes. Initial ROP/NOP flux shapes are based on an idealized/stylized time-average (ripple-free) power distribution, which simplifies the analysis – this is a quasi-steady-state approach. The effect of ripple is then accounted for subsequently through ROP/NOP detector calibration while the reactor operates.

In the real core, the instantaneous refuelling ripple may result in the vulnerable channel (for a particular upset) having a higher instantaneous power than the reference time-average power. The worst-case situation would be the one where the vulnerable channel is the one with the highest ripple, which is the CPPF. This difficulty is solved by recognizing that the actual detector readings don't matter, only the MTT matters. The ROP/NOP detectors do not need to read 100% with the reactor at full power. Instead, all ROP/NOP detectors are calibrated to the product of the CPPF and the current reactor power.

Detector calibration is done at every shift, while operating in a reasonable steady state. Detector calibration with CPPF preserves the MTT for the worst-case scenario (i.e., the MTT is decreased by CPPF for every detector). Trips will occur early if the vulnerable channel is not the CPPF channel. To maximize the MTT, the CPPF must be kept as low as possible. This requires a careful and judicious selection of channels for refuelling, and/or the use of different axial refuelling schemes in different regions. Determining the CPPF value, and ensuring detectors are calibrated to the correct value, are on-going duties of the fuelling engineer or the station physicist.

6.5 Treatment of Uncertainties

The uncertainties considered in the ROP/NOP analysis cover CCP uncertainty, channel power uncertainty, ROP/NOP detector-related uncertainty, and flux-shape-related uncertainty. The uncertainties are further classified as to whether or not they are specific to each channel/detector or are common to all channels/detectors:

- detector specific (random error which varies independently from detector to detector);
- channel specific (random error which varies independently from channel to channel);
- detector common (random error common to all detectors); and
- channel common (random error common to all channels).

The CANDU community currently uses SIMBRASS and ROVER-F as active licensing analysis tools for ROP/NOP TSP calculations. In the traditional ROP/NOP analysis method with ROVER-F, uncertainties of a similar character, such as channel-random terms that affect the MTD, are combined into the following four primary uncertainty categories, to simplify their application: channel-random, detector-random, common-random (channel and detector terms), and systematic biases. In the latest version of SIMBRASS, uncertainties are classified as aleatory and epistemic, and the subsequent treatment of each uncertainty value depends on this classification.

6.6   The Impact of PHTS Aging on ROP/NOP TSP

The PHTS operating conditions (coolant flows, temperatures, pressures) in a CANDU reactor are affected by the aging of the PHTS components (pressure tubes, SGs and feeders). Over time, this results in less effective fuel cooling and consequently lowers the power at which the fuel will experience dryout. Certain corrective measures, such as mechanical and/or chemical cleaning of the PHTS components, design changes (e.g., implementation of the 37M fuel), and/or other operating measures, could significantly slow down this trend, but, eventually, adjustments to the ROP/NOP TSP might be needed. The lowering of the ROP/NOP TSP would eventually lead to a reduction (derating) in the electrical output allowed at the station. Hence the ROP/NOP TSP analysis for the slow LOR event needs to incorporate the impact of PHTS aging.

7. Short-Term Reactivity Change: Xenon Effects

7.1 Introduction

Most fission products absorb neutrons to some extent and they accumulate slowly as the fuel burnup increases, hence decrease the long-term reactivity. The neutron-absorbing fission-product xenon-135 has particular operational importance. Its concentrations can change quickly in a power maneuvre, producing major changes in neutron absorption on a relatively short-time scale (minutes).

We start by considering the mechanisms for creation and elimination of Xe-135 and its precursor, iodine-135. This allows us to derive expressions for their steady-state concentrations. It also enables us to analyze the sequence of events following a power reduction or shutdown after a prolonged operation at power, and to understand why this leads to a rapid increase in Xe-135 concentration.

Any power manoeuvre produces a transient change in xenon concentration. Reactivity changes caused by xenon-concentration changes are not as immediate as the reactivity changes caused by a change in fuel temperature, but the size of the effect can be much larger. Immediate feedback from xenon is positive. For example, a power increase causes an increase in reactivity that tends to cause a further power increase. This is important, because the reactivity effect can exceed the capability of the regulating system to compensate for it.

Another important consequence of the presence of Xe-135 in a CANDU reactor is that it leads to the possibility of spatial xenon oscillations. These can cause power in some core regions to rise and fall, with a period of 15-30 hours, with the possibility of overrating the fuel. We will describe the process by which spatial xenon oscillations can occur, and see why they necessitate continuous flux monitoring at a number of points in the reactor.

7.2 Xenon Reactivity Effect

Xenon-135 (often simply referred to just as xenon) is the most important fission product. It has a very large thermal absorption cross section (~2.66×10^6 barns). Xe-135 is produced in two ways:

- Directly from fission: about 0.6% of all fissions in an equilibrium CANDU fuel produce Xe-135;
- Indirectly from the decay of I-135 (which has a half-life of ~6.6 hours). I-135 is produced either directly in fission, or as a daughter of the fission-product tellurium-135. About 6.4% of all fissions in CANDU fuel produce either I-135 or Te-135. Due to the short half-life (~19 seconds) of Te-135, we can simply take ~6.4% as the I-135 yield.

Xe-135 is also destroyed in two ways: through its own radioactive decay (which has a half-life of ~9.1 hours), and by absorption of neutrons to form Xe-136 (this pathway is called burnout). At high power, neutron capture removes much more xenon than does beta decay. An important point is that the burnout rate changes immediately when flux changes, while the beta decay of Xe-135 is governed by the 9.1-hour half-life. The production and loss mechanisms for I-135 and Xe-135 are shown schematically in Figure 13.

Xe-135 is a strong neutron absorber so its presence in the fuel creates a large negative reactivity in the core. The reactivity worth of the Xe-135 is known as the "xenon load". At steady-state full power, the reactivity worth of equilibrium xenon in a large CANDU is about -28 mk. The Xe-135 reactivity at steady-state lower-power levels (down to about 50%) is only a couple of mk lower.

It is also common practice to express the concentration of I-135 as the "iodine load" (in mk). It is important to realize that iodine itself is not a significant poison; there is no appreciable reactivity associated with it. The reason for the definition of "iodine load" is that I-135 can be considered as a "reservoir" of Xe-135, so the iodine load represents the reactivity that would be inserted into the reactor if all the iodine suddenly changed into xenon. Note this point carefully, that we are not talking about a real iodine reactivity that exists in the system, but an

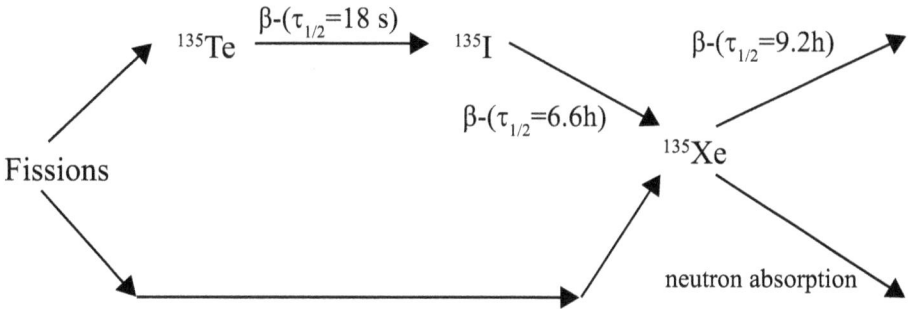

Figure 13   Production and Loss Mechanisms for I-135 and Xe-135

"iodine reactivity reserve bank" of potential reactivity that could gradually show up in the form of neutron-absorbing Xe-135. For a large CANDU, the equilibrium "iodine load" (or "iodine reactivity reserve bank") at full power is worth about -320 mk. The value for any particular reactor depends on the value of the steady-state flux at full power.

7.3   Iodine and Xenon: Equilibrium Buildup

Xe-135/I-135 kinetics leads to the saturating character of iodine and xenon. Figure 14 shows the "iodine reactivity reserve bank" (iodine load) in fresh fuel grows to within 2% of its equilibrium value after about 40 hours of residence in reactor. Note that the equilibrium level of I-135 is directly proportional to the thermal-neutron flux (or power). Figure 15 shows that the buildup of negative Xe-135 reactivity (xenon load) to its equilibrium value of -28 mk takes also about 40 hours; the upper curve is for the buildup at full power (P = 1), the lower curve is for buildup at 60% full power (P = 0.6).

The equilibrium negative Xe-135 reactivity (xenon load) shown in Figure 15 is nearly the same at 60% and 100% full power. This is quite different for the equilibrium "iodine reactivity reserve bank" (iodine load), which is directly proportional to the power level. At high flux, where the xenon burnout rate is much greater than the decay rate, the equilibrium xenon load becomes nearly independent of flux. This condition applies within the range of about 60% to 100% of full power in CANDU.

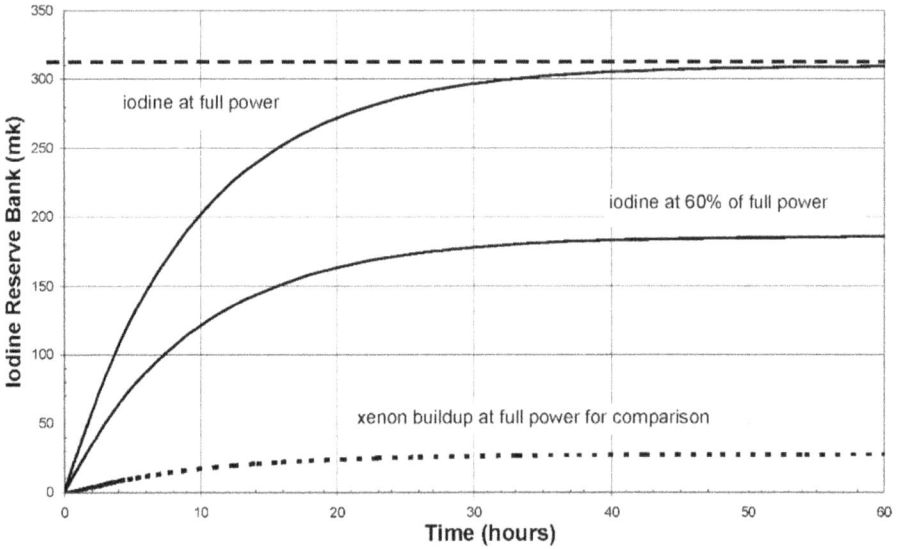

Figure 14　Buildup of Iodine Reactivity Reserve Bank (Iodine Load) to its Equilibrium Value

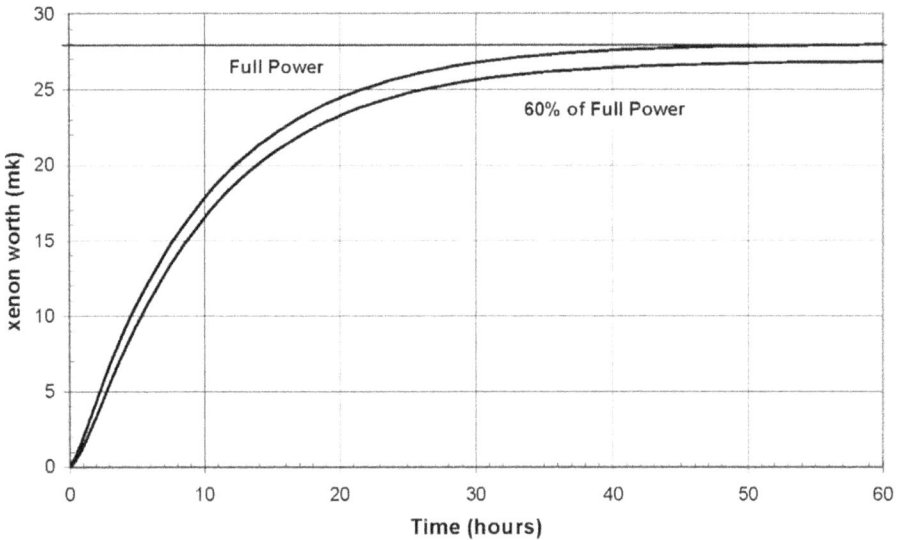

Figure 15　Buildup of Negative Xe-135 Reactivity (Xenon Load) to its Equilibrium Value

7.4 Transient Xenon Behaviour

Although the reactor has sufficient excess reactivity to offset equilibrium xenon, a problem occurs following a shutdown after operation at power. The xenon load rises (becomes more negative) rapidly, peaking in about 10 hours, after which it decreases again, slowly. It takes nearly two days for the xenon load to return to near the full-power equilibrium value. During this time, there is not enough excess reactivity available to make the reactor critical, so it remains shut down.

For a CANDU that has been running steadily at full power long enough to establish equilibrium xenon conditions, the production rate of xenon from iodine decay is much larger than its production rate directly from fission, the ratio is about 9:1. The removal rate of xenon by burnout is much larger than its removal rate by its own radioactive decay, again by roughly 9:1.

Just before shutdown, xenon removal (primarily by burnout) matches the xenon production (mainly from iodine decay). Following a reactor shutdown, the flux drops to a near-zero value within a minute or so, stopping direct xenon production from fission almost immediately, but xenon production by iodine decay continues. Net production remains near 90% of the equilibrium value. On the loss side, we lose 90% of xenon removal, as the xenon burnout drops to zero, leaving only radioactive decay. The result of these two effects is that the xenon load starts to rise (becomes more negative) quite rapidly, fed by the decay of iodine, as shown in Figure 16.

This cannot continue indefinitely, because there is a limited quantity of iodine in the core and iodine production by fission stops following reactor shutdown. Iodine decays with its characteristic half-life (~6.7 hours). The xenon concentration therefore reaches a peak value, about 10 hours after shutdown. Thereafter the xenon load gradually decreases, as its reduced rate of production by iodine decay cannot keep up with the enhanced xenon decay.

7.5 Poison Override

One function of the adjuster rods is to provide excess reactivity to override xenon transients. Withdrawing adjuster rods from the reactor core

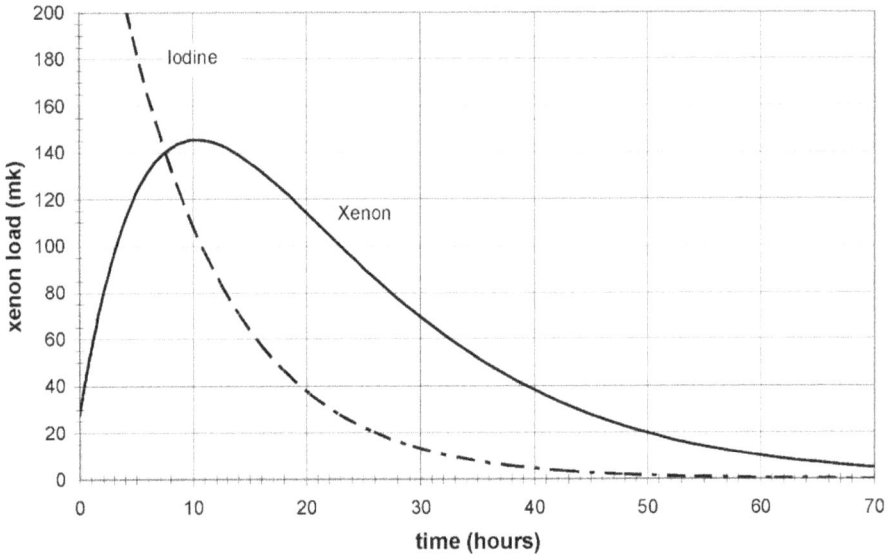

Figure 16   Change in Negative Xe-135 Reactivity (Xenon Load) Following a Trip from Full Power for an Ontario CANDU (*Source:* Figure 5.5 from [4])

contributes positive reactivity, up to a maximum depending on the particular reactor, about 15 mk in CANDU 6. If the negative reactivity due to xenon exceeds the adjuster reactivity worth, the reactor is subcritical, with no way to restart it. We say the reactor has poisoned out. Figure 17 shows that on a reactor trip from full power, a reactor that has poisoned out cannot be restarted until about 35 to 40 hours after the trip, when xenon has decayed back to near the -28 mk equilibrium level.

Holding reactor power near 60% (or higher) allows sufficient xenon burnout to prevent a "poison-out" (as Figure 17 shows). It is important to realize that on a turbine trip it may make economic sense to keep the reactor operating by exhausting steam to the condenser (or the atmosphere), to prevent poison-out and allow a quick return to higher power when possible. We call this mode of operation "poison-prevent".

The rate of rise of the xenon load after a trip is a function of the equilibrium conditions before the trip. In CANDU reactors, the xenon load increases (becomes more negative) at about 0.5 mk per minute

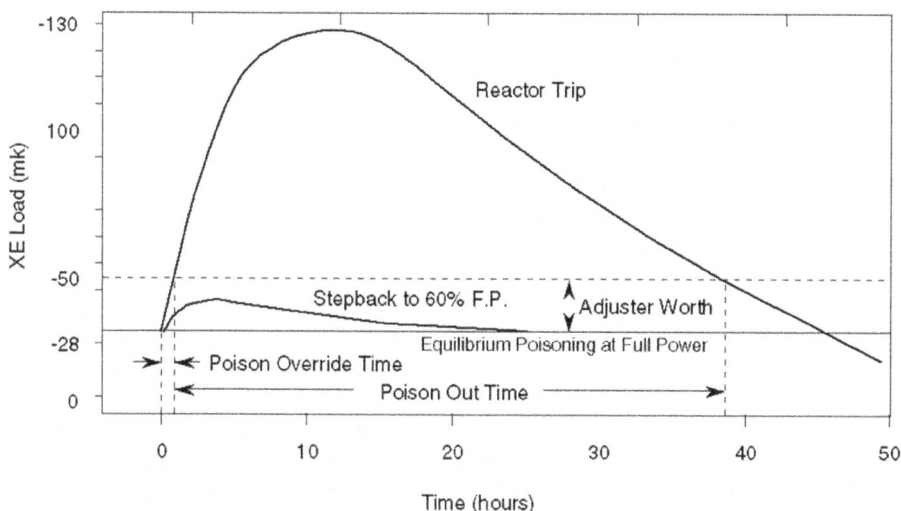

Figure 17 Change in Negative Xe-135 Reactivity (Xenon Load) Following a Trip from Full Power and a Setback from Full Power to 60% FP in CANDU 6 (*Source*: [10])

following a trip from full power. This number and the available reactivity for poison override determine the poison-override time (the time from the trip to a "poison-out"). For example, if a particular reactor has a maximum available reactivity of 15 mk from adjuster rods, it must be brought back to high power within about 30 minutes (15 mk/~0.5 mk/min), because after this time the negative reactivity from xenon exceeds the reactivity worth of the adjusters (the poison-override capability).

Poison override is possible in principle, and is part of the reactor design, but is often not practical. Before restarting the reactor following a trip, it is important to find the cause of the trip and eliminate the fault. The operators must make a number of checks before judging the trip to have been spurious. Checks following a trip (or repairs) can often take longer than the time required to prevent "poison-out".

7.6 Xenon Transients Following Power Changes

In practical reactor operation, we are also interested in the transients following power changes different from a shutdown. Figure 18 shows the transients for power reductions by 20, 40, 50, 60, 80, and 100% from initial full power. For a reduction of, say, 40% (that is, from 100% to 60% power) the xenon removal by burnout also decreases by 40% from its full-power value, but because there is still sufficient burnout, the transient will not result in the xenon concentration increasing to its peak value in a shutdown. The figure shows that for a reduction to 60% FP, an available excess reactivity of ~15 mk would be sufficient to override the transient altogether. Ultimately, iodine reaches a new equilibrium at 60% of its full-power value, and the xenon reaches equilibrium with the xenon load a little less than -28 mk.

It is interesting to note that the converse of these curves applies to increases in power from equilibrium operation at lower values of power to 100% power. Normally, xenon transients on power increases do not

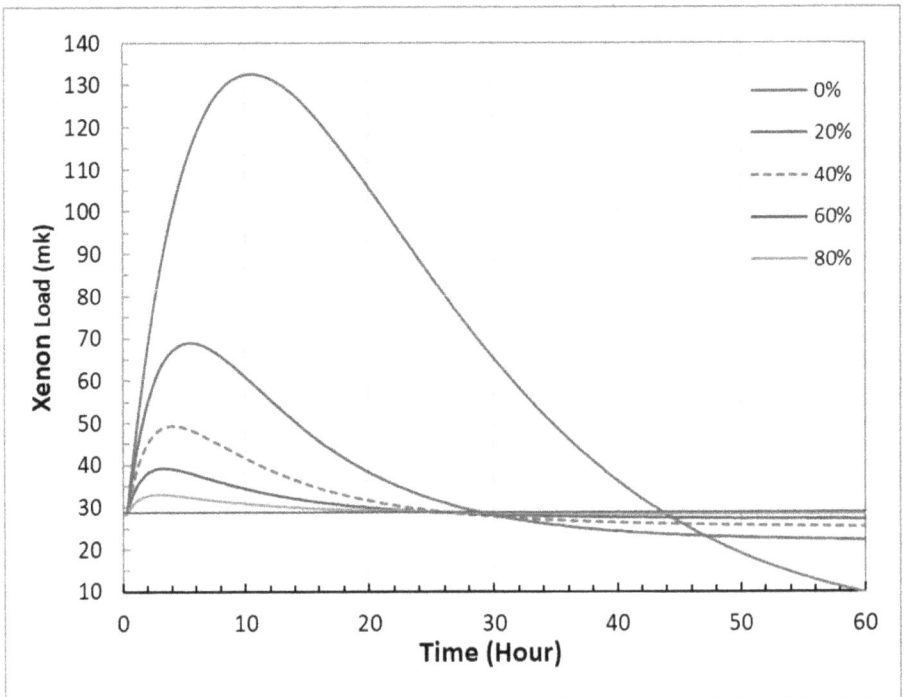

Figure 18 Change in Negative Xe-135 Reactivity (Xenon Load) Following Step Power Reductions from Equilibrium Full Power in CANDU 6

present particular operational problems, as the addition of neutron-absorbing poison (boron or gadolinium) to the moderator can offset the excess reactivity resulting from the rapid decrease in xenon load (as it becomes less negative).

7.7   Spatial Xenon Oscillations

A change in local flux causes a rapid change in xenon burnout, while the change in xenon production is delayed by iodine buildup and decay. This can cause a reactor to undergo periodic (i.e., repetitive) oscillations in flux level. A suitably designed control system limits the size of the resulting flux peaks and stabilizes the flux.

For a small reactor, careful monitoring of bulk reactor power permits remedial reactor power regulation and prevents oscillations. For a large reactor, however, such as a CANDU, monitoring only the overall power level is inadequate, because local xenon oscillations can drive power up in one region of the reactor and down in another, even though the total power is maintained constant. In fact the potential for localized xenon oscillations is somewhat higher in CANDU than in other reactors on account of daily local perturbations (online refuellings), as described below.

Suppose that the reactor is operating at high power with a symmetric side-to-side power distribution, then a power tilt may develop after a few channels on the left side are refuelled, without refuelling a similar number on the right side. Now, suppose that the regulating system continues to keep the total power output of the reactor constant, but spatial control is lost or is inadequate to fight the asymmetric change in power. The flux then increases a little on the left side of the core, and simultaneously decreases on the right side. The scenario described in this particular example is a side-to-side flux and power tilt.

In the region of increased flux, xenon now burns out more rapidly than it did before the change and its concentration decreases. The decrease in xenon concentration leads to a higher flux, which again results in increased local xenon burnout, increased local reactivity, increased flux and so on.

Meanwhile, in the region of decreased flux, the xenon concentration increases on account of its reduced burnout and to the continued decay of the existing iodine produced in the original higher flux. This increased xenon concentration decreases reactivity in this region, which reduces the flux, and in turn increases the xenon concentration, and so on. The flux, and hence the power density, decreases in this region and increases in the other, while the total power of the reactor remains constant.

These opposite local power excursions do not continue in the same direction forever. At the same time as the increased flux is causing xenon to burn out more rapidly in the high-flux region, it is also increasing the production of iodine. The decay of this enhanced iodine bank eventually leads to an increase in xenon concentration, reducing reactivity and thus the flux and power in that region. Likewise, in the region of reduced flux, the lowered production of iodine combined with the decay of accumulated xenon increases the local reactivity and reverses the flux and power transient in that region.

In this way, unless the regulation system responds adequately to control these local transients, the flux and power may oscillate between different regions (top-to-bottom, end-to-end, or side-to-side) indefinitely. Calculations show that xenon spatial oscillations have a peak-to-peak cycle time of about 20 to 30 hours, and the height of the peaks may increase from cycle to cycle.

7.8   Conditions for Spatial Oscillations

The type of localized xenon oscillation described above can take place only in a large reactor, a reactor whose spatial dimensions are large compared to the diffusion length of the neutrons. With a small core, a disturbance started in one region affects other regions very quickly, because neutrons from the affected region spread the disturbance quickly across the core. As mentioned earlier, a regulating system that controls bulk power adequately prevents oscillations in a small reactor.

When the dimensions of the reactor greatly exceed the distance travelled by the thermal neutrons during their lifetime (which is the case in a large CANDU), a disturbance that begins in one place does not spread its influence quickly to a distant part of the core, so the various regions act much more independently. Thus, if a flux increase occurs in one region because of a refuelling, for example, a control system based on maintaining the overall power constant will reduce the flux a little throughout the core to compensate. This would set up a xenon oscillation in the second region exactly out of phase with the one in the first region.

The other condition that must pertain before spatial xenon oscillations can occur is that the reactor is operating at high power. When the flux increases at some point in the reactor, the immediate increase in xenon burnout initiates the oscillation. For a marked change in xenon concentration, xenon burnout must predominate over xenon decay. We have already seen that this is the case for a large CANDU, where xenon burnout at full power is at least a factor of 9 larger than xenon decay. Typically, a spatial disturbance cannot become an oscillation in a CANDU unless the reactor power is above 25% full power.

The CANDU, then, like several other types of power reactors, satisfies the two conditions required for spatial xenon oscillations to occur. Oscillations can occur with constant (overall) reactor power, so they can continue unnoticed unless instruments monitor the flux and/or power at several points throughout the reactor, and localized absorbers respond to adjust local reactivity to counteract any flux tilts.

7.9   Limiting Spatial Oscillations

All CANDU designs use 14 LZCs coupled with 14 pairs of in-core detectors to detect and adjust the liquid-zone levels to provide spatial control and counteract xenon transients, in particular spatial oscillations. Even with a zone control system, a severe xenon oscillation could risk significant damage to the fuel. The size of the flux tilt depends on the size of the reactivity upset that initiates the oscillation, and on the amount of compensating reactivity that the LZC system provides. A large oscillation could drive one or more liquid-zone levels to their operating limit and spatial control would be lost in these zones. Continued operation with a flux oscillation of such magnitude could lead, at least, to a reactor trip or, more seriously, to dangerously high local fuel temperatures or even fuel meltdown. Even without such severe consequences, xenon oscillations could burden the core materials with unnecessary temperature cycling that could lead to premature materials failure.

7.10   Samarium

Samarium-149 (often simply referred to just as samarium) has a large thermal absorption cross section ($\sim$4.05$\times$10^4 barns) and a high production rate (total fission yield is $\sim$1.2%). Neither the cross section nor the yield are as big as for Xe-135, so the samarium reactivity effects are much smaller. Fission actually does not produce Sm-149 directly. Sm-149 is the daughter of neodymium-149 and promethium-149. Because of the short half-life of Nd-149 ($\sim$1.7 hours) compared to the half-life of Pm-149 ($\sim$53 hours), we consider the entire $\sim$1.2% fission product yield to be promethium.

Like I-135, Pm-149 does not absorb neutrons; only decay can remove it. Promethium buildup, therefore, is similar to iodine buildup. Because of its long half-life, it takes about 300 hours to reach equilibrium, compared to 40 hours for I-135. As with iodine, equilibrium promethium concentration is proportional to flux. The expression for Sm-149 buildup is simpler than for Xe-135, because there is no direct production of samarium from fission and no loss by decay.

Unlike xenon, the time required to reach equilibrium is a function of the flux level, but the equilibrium samarium concentration is independent of the flux (for all power levels). Another important difference from xenon is that samarium-149 is a stable nuclide and does not decay after shutdown, so these samarium buildup curves, with no initial samarium, apply only to fresh fuel inserted in the reactor.

After a reactor shutdown, there is a transient rise in samarium concentration because promethium decay continues, but burnout by neutron capture ceases when the flux disappears. The maximum samarium load after shutdown depends on the promethium load before shutdown, which depends on the reactor flux. For CANDU reactors, the maximum samarium load after shutdown is about 9-12 mk. Although reactor design must allow for the equilibrium samarium load, the shutdown load does not cause operational problems for the following two reasons:

- The maximum samarium load appears long after the xenon peak decays. There will be lots of reactivity available to deal with the samarium buildup when it occurs. The transient rise in samarium is negligible during the xenon-override time, so this does not present a problem either.

- There is a plutonium-239 transient buildup (from Np-239 decay) that increases reactivity by a similar amount and at about the same rate as samarium decreases it. The rate at which samarium is formed after shutdown is governed by the Pm-149 half-life of ~53 hours which, by coincidence, is almost the same as the ~56-hour half-life of Np-239. After shutdown, the Pu-239 starts to increase above its pre-shutdown value because Np-239 decay continues, but Pu-239 burnout stops. It turns out that the increased reactivity gained from the Pu-239 buildup more than offsets the reactivity loss due to the increased Sm-149. The net result may be a small reactivity gain of a couple of mk.

Although samarium does not decay during shutdown, it will burn back to equilibrium following a return to power. On restart, the promethium load builds to equilibrium over ~300 hours just as described earlier. The burnout rate of samarium is significantly faster than this. It is also faster than the reduction of excess Pu-239 to equilibrium.

The net effect is that, after xenon returns to equilibrium, there is an excess reactivity of several mk that disappears in a few more days. This does not cause operational problems, unless normal refuelling stops (in the belief that there is enough excess reactivity). If this happens, the core will require rapid refuelling when the excess reactivity vanishes with the excess plutonium.

Similarly, transient changes in samarium on power-level changes are very small compared to xenon, and change very slowly over a week or so. In each case, the samarium level returns to the same equilibrium value and the LZC system easily correct for small deviations from this value.

8. Short-Term Reactivity Change: Local-Parameter Reactivity Feedback

8.1 Local Parameters Affecting Reactivity in a CANDU

A change in power for an operating reactor generally alters local parameters in the reactor such as the temperatures of the fuel, moderator, and coolant. A change in any of these local parameters causes a change in reactivity that, in turn, affects reactor operation (a feedback effect). Local parameters help to understand the feedback reactivity components related to the core evolution. For the CANDU reactor, the most important local parameters are the following:

1. Fuel Burnup — (0 - 8000 MWd/Mg(U))
2. Neutron Flux — ($\sim 10^{14}$ at full power)
3. Fuel Temperature (FT) — (25 - 1500 °C)
4. Coolant Temperature (CT) — (25 - 310 °C)
5. Coolant Density (CD) — (0 - 1.1 g/cm^3)
6. Coolant Purity (CP) — (< Moderator Purity)
7. Moderator Temperature (MT) — (25 - 100 °C)
8. Moderator Density (MD) — (0 - 1.1 g/cm^3)
9. Moderator Purity (MP) — (99.75 wt% D$_2$O - 99.96 wt% D$_2$O)
10. Moderator Boron (MB) — (0 - 1 ppm; 90 ppm for GSS)
11. Moderator Gadolinium (MGD) — (0 - 3 ppm; 16 - 25 ppm for GSS)
12. Moderator level — (applicable to Pickering CANDU units only)

Strictly speaking, parameters 1 to 5 are true local parameters as they are distributed unevenly in the CANDU reactor core due to the neutron power distribution, the thermalhydraulic conditions in the fuel channel, and the refuelling strategy. Parameters 7 to 12 are almost uniformly distributed in the CANDU reactor core and they are generally considered independent of the neutronic power, hence they should be called "global" parameters rather than "local" parameters. But for simplicity, they are all treated as local parameters in this section.

Fuel burnup is distributed unevenly in the CANDU reactor core due to the power distribution and the refuelling strategy. The nuclear properties of a fuel bundle (also called lattice-homogenized macroscopic cross sections, or lattice cross sections) in CANDU are dependent on the fuel isotopic composition, which changes with fuel burnup. The most important local parameter is thus the fuel burnup of individual bundles. With on-line refuelling, each fuel bundle in a CANDU reactor has its own specific power history, leading to a specific fuel composition. The effect of fuel burnup and of its history on lattice cross sections and on the power distribution must therefore be taken into account (see Section 11.5.5).

The neutron flux affects directly the concentration of the poisons xenon and samarium. Reactivity loads are associated with their concentrations as described in Section 7. The neutron flux also affects the fuel temperature, coolant temperature and coolant density, as discussed in detail in the following sections.

8.2 Reactivity Feedback and Reactivity Coefficients

The reactivity feedbacks caused by changes in the local parameters are quantified as reactivity coefficients. A reactivity coefficient for a given parameter gives the instantaneous or near-instantaneous change in reactivity per unit change in the parameter. For instance, the temperature coefficient of reactivity is the change in reactivity per unit increase in temperature. Its units are mk/°C or pcm/°C.

Temperature changes occur in the fuel, coolant, and moderator more or less independently so there is a temperature coefficient of reactivity associated with each. The reactivity effect of a temperature change is (to first order) the product of the temperature reactivity coefficient and the temperature change. It is helpful to know typical temperatures of the fuel, moderator, and coolant in different operating states. Table 5 shows the typical effective average temperatures of fuel, coolant, and moderator

Table 5 Typical Temperatures (°C) of Reactor Components

Component	Cold Shutdown	Hot Shutdown	Full Power
Fuel	25	290	690
Coolant	25	265	290
Moderator	25	66	70

for CANDU 6 at cold shutdown, hot shutdown and full power. The term "cold shutdown" means PHTS is at atmospheric pressure and room temperature, and "hot shutdown" means PHTS is pressurized (10 MPa) and at operating temperature. Temperatures in particular stations may differ somewhat from these values. Table 6 shows important reactivity coefficients for the CANDU reactor.

Note that the safety importance of an individual reactivity coefficient is a function of its magnitude, as well as the speed of the parameter change and its sign (positive or negative). The reactor characteristics that impose least demand on the control and protective system appear to be those in which the individual reactivity coefficients are small. Usually it is desirable for a reactor to have negative reactivity feedbacks to provide self-regulating features. In particular, it is helpful if the fuel temperature coefficient is negative, because in an

Table 6 Typical Values of CANDU Reactivity Coefficients

Reactivity Coefficients	Value
Coolant-Void Reactivity	-15 to -20 mk
Power Coefficient of Reactivity	close to zero (positive)
Fuel-Temperature Coefficient of Reactivity	close to zero (negative)
Coolant-Temperature Coefficient of Reactivity	Small positive
Moderator-Temperature Coefficient of Reactivity	Small positive
Moderator-Purity Coefficient of Reactivity	~ 33 mk/at%
Coolant-Purity Coefficient of Reactivity	~ 3 mk/at%
Moderator-Boron Coefficient of Reactivity	-~ 8 mk/ppm
Moderator-Gadolinium Coefficient of Reactivity	-~ 28 mk/ppm

increasing-power transient the fuel heats up more rapidly than the other core components (such as coolant) do.

For those countries where only, or predominantly, LWR technology is deployed, it was found that regulations tend to include references to negative reactivity feedback characteristics. In contrast, international regulatory approaches, such as those of the International Atomic Energy Agency, and which encompass a range of reactor technologies, do not emphasize negative reactivity feedback requirements, but rather emphasize defense in depth, as well as the requirement to show that all initiating events covering operating transients and accidents are safely mitigated. The CANDU safety case is based on small values of reactivity coefficients and total reactivity change, coupled with dual, redundant, fast-acting independent SDSs. The regulator sets performance objectives related to effectiveness of systems to control and to shut down the reactor, and to the reliability of the control and shutdown functions.

8.3 Fuel-Temperature Coefficient of Reactivity

The fuel-temperature coefficient arises principally from two factors: one is the Doppler broadening of resonances, and the other is due to the change in the neutron spectrum with temperature. Doppler broadening reduces the resonance-escape probability (primarily in U-238) and this is much larger than all other reactivity effects in the fuel. The spatial variation of fuel temperature can be taken into account by calculating the effective fuel temperature as a function of power density. The effective fuel temperature is lower than the volume-average fuel temperature, since the neutron-flux distribution is non-uniform through the fuel pellet and this gives preferential weight to the surface temperature.

The reactivity effect due to a change in the fuel temperature is the fastest reactivity effect. It appears within fractions of a second. In CANDU it is a small and negative component. It becomes less negative with fuel burnup, on account of the 0.3-eV fission resonance in Pu-239. It can change sign from negative to positive at a high fuel burnup, but the average value in the equilibrium core is negative.

Under normal operations, the fuel temperature is limited by the maximum permissible channel power, bundle power, and fuel-element power. The average fuel temperature in a fuel element is often around 690 °C. Centreline temperatures are much higher. In a guaranteed shutdown state (GSS, see Section 10.3), the fuel temperature is somewhat higher than ambient temperature because of the decay heat of fission products within the fuel. Under some accident conditions the fuel temperature will increase by 50 °C to 100 °C before the shutdown systems respond. In some LOCAs being studied, the temperature rise may reach several hundred °C.

8.4 Coolant-Void Reactivity

The CANDU Coolant-Void Reactivity (CVR) is defined as the reactivity change due to a complete loss of coolant from the core. It is positive throughout the fuel lifetime in the core. The root cause lies in the fact that CANDU is a pressure-tube reactor, with the coolant separated from the moderator. In LWRs, one liquid serves as both coolant and moderator, and a loss of coolant is also a loss of moderator, leading to a decrease in reactivity. In CANDU, however, the loss of coolant does not imply a significant reduction in moderation, which is provided in great part by the moderator outside the pressure tubes. Overall, the particular effects that lead to the positive CVR in CANDU are a reduction in resonance capture of neutrons and an increase in the fast-fission rate.

As the fuel is irradiated, plutonium and fission products build up, and the change in neutron spectrum gives a negative component in the CVR. This is due to a reduction in absorptions in the 0.3-eV fission resonance of Pu-239, i.e., a reduction in fissions from that source. However, the net reactivity change on coolant voiding is still positive (but smaller than for fresh fuel).

The magnitude of the CVR is sensitive to the concentration of absorbers in the lattice cell. The two most important of these are moderator poison (boron or gadolinium) and ordinary water in the heavy-water coolant (i.e., downgraded coolant purity). In general, the CVR becomes more positive with degraded coolant purity and increased moderator poison concentration (the latter, on account of the hardening of the thermal-neutron spectrum).

Full-core CVR ranges from 15 mk to 20 mk, depending on the fuel burnup and other parameters. Note that the actual full-core CVR calculated with the 3D core model used in the accident transient analysis is generally higher that the lattice CVR calculated from single-lattice calculations, because the change in leakage and the 3D spatial distribution of the coolant and flux shape are properly accounted for in a full-core calculation. Note that the size of the CVR is more than sufficient to reach prompt criticality if the coolant is fully lost very quickly and SDSs do not act in a timely manner.

A Large LOCA (LLOCA), caused by the rupture of a large pipe, such as a Reactor Inlet Header (RIH), a Reactor Outlet Header (ROH), or a pump-suction pipe, is a hypothetical accident which must be analyzed. The amount of voiding in a postulated LOCA depends on the size of the break. Fortunately, in CANDU, which has two coolant loops side by side and bidirectional coolant flow, the coolant voiding occurs quickly in only one loop and in only of the passes in that loop, and it takes time for all the coolant to flash to steam through a rupture; long enough for emergency shutdown instrumentation to detect the power increase and trigger a reactor trip. In CANDU, a LLOCA can inject ~4-5 mk of positive reactivity in the first second after the break, which is beyond the capability of the RRS to control. Thus a LLOCA leads to a sudden power surge (power pulse), which must be terminated by a SDS. The most likely reactor trips, which will lead to SDS actuation, are a high log rate measured by SDS ion-chambers, or high reactor power measured by in-core detectors.

A LLOCA is in fact the accident which presents the greatest challenge to CANDU shutdown systems in terms of the rate of positive reactivity insertion. This DBA has provided the impetus to provide two independent SDS (SDS1 & SDS2) in CANDU technology.

8.5   Coolant-Temperature Coefficient of Reactivity

The reactivity effect due to a change in coolant temperature appears within seconds. In fresh CANDU fuel the coolant-temperature coefficient, excluding density effects, is expected to be negative. As burnup proceeds, a shift from a small negative to a small positive coefficient is expected, again due to the 0.3-eV fission resonance in Pu-239. For equilibrium fuel, the coolant-temperature coefficient excluding the density effect is positive throughout the whole temperature range. Its value near the normal operating temperature is typically 0.02 mk/°C.

The coolant temperature inside a CANDU pressure tube may vary anywhere from room temperature (about 25 °C) in the initial startup of the reactor to about 300 °C at full power. During startup after a long shutdown, the coolant temperature can begin at 80 °C. Under normal operating conditions at full power, the temperature ranges from 260 °C to 310 °C; and the density of the heavy-water coolant lies between 0.75 and 0.88 g/cm³.

8.6   Power Coefficient of Reactivity

The power coefficient of reactivity (PCR) is an overall measure of the reactivity change per unit increase in reactor power, typically quoted in mk/%FP. It can be evaluated for any power level, but tends to be of greater interest at high-power operation. When there is a change to the operating power of a reactor, physical properties (fuel temperature, coolant density, etc.) will change. Hence, the PCR is a combination of several feedback mechanisms, including fuel temperature, coolant temperature, and coolant density. In CANDU the contribution from coolant void is of greatest interest, while for LWR it is the contributions from fuel temperature and coolant temperature.

The PCR value depends on the starting power level and on operating conditions such as fuel burnup, fuel and coolant temperatures, and extent of coolant boiling/subcooling in channels. For LWR designs, given the large negative values of fuel-temperature and coolant-temperature coefficients, the PCR is negative. Typical values of the PCR of Pressurized-Water Reactors (PWRs) are about -0.11 mk/%FP at the beginning of cycle and about -0.23 mk/%FP at the end of cycle in the power operating range. The CANDU core, on the other hand, has very low values of PCR for all conditions (about 0.02 mk/%FP at full power). The PCR could be positive or negative, depending on the core condition). Whether the PCR is positive or negative, the small magnitude of the PCR leads to relatively small changes in core conditions with time (power manoeuvering from 0 to 100% full power requires very little adjustment of reactivity devices). The CANDU PCR has little impact on operation or safety, on account of the long CANDU prompt-neutron lifetime. The small magnitude of the PCR and the long neutron lifetime contribute to a stable, readily controlled CANDU core.

8.7 Moderator-Purity Coefficient of Reactivity

The neutron-absorption cross section of hydrogen (H-1) is orders of magnitude higher than that of deuterium (D, H-2). The moderator isotopic purity has a direct and strong impact on reactivity. Under equilibrium-fuel conditions, the reactivity decreases by about 33 mk per percent increase in the moderator H_2O isotopic concentration. Hence, under normal operations, the moderator purity is kept as high as possible, to maximize the fuel discharge burnup. The moderator purity is almost always maintained above 99.90 wt% D_2O in CANDU plants, and 99.95 wt% D_2O is not uncommon.

8.8 Moderator-Poison Coefficient of Reactivity

Moderator poison is primarily used for reactivity shim (i.e., to suppress excess reactivity) in the following conditions:

1. Reactor startup at the cold condition with all fresh fuel without xenon;
2. Reactor fuelling-ahead to "store" up excess reactivity;
3. Reactor restart after a long shutdown, when the saturating fission products have decayed;
4. Reactor trip with SDS2; and
5. Reactor in a GSS.

In general, boron is used in the former two situations and gadolinium in the latter three situations. Gadolinium has a large neutron-absorption cross section than boron: 1 ppm Gd is worth about 3.5 ppm B under cold reactor conditions, and 3.4 under hot operating conditions. The burnout rate of gadolinium on a startup after a shutdown is comparable to the xenon growth rate, hence smooth control is possible with gadolinium. Gadolinium is the preferred moderator poison to achieve a poison-based GSS (see Section 10.3).

During normal operations the moderator poison levels are close to 0.0 ppm (some residual amount is difficult to extract). During normal operational manoeuvres (e.g., fuelling ahead, restart after long shutdown), moderator poison levels are adjusted within the ranges of 0-11 ppm of boron and 0-3 ppm of gadolinium, for example:

- 6 ppm boron for reactor restart with adjusters withdrawn in CANDU-6 reactor;
- 1 ppm boron for fuelling ahead by 8 mk; and
- 4 ppm boron at the end of the outage for the equilibrium core. Following startup, as reactor power is increased, adjuster rods would be reinserted and fission-product inventories would increase, requiring extraction of the moderator poison.

8.9   Importance of Reactivity Coefficients in Safety Analysis

Knowledge of the various reactivity coefficients is important from the point of view of reactor control and safety. The RRS must be designed to respond to small perturbations in core parameters, and two independent, highly reliable SDSs must be able to shut down the reactor for any postulated DBAs such as LLOCA, small LOCA (SLOCA), in-core LOCA, steam-line break (SLB), slow LOR, fast LOR, loss of flow (LOF), loss of moderator inventory (LOMI), etc. Table 7 summarizes the importance of individual reactivity coefficients on the safety analysis of these DBAs, and on the physics design. Note that a reactivity coefficient is marked √ if it has a direct and significant impact on a specific parameter, or if it directly

Table 7 Importance of Individual Reactivity Coefficients on CANDU Safety Analysis and on Physics Design

No	Reactivity Coefficient	Safety Cases									Physics Design		
		LLOCA	SLOCA	In-Core LOCA	SLB	slow LOR	fast LOR	LOF	LOM	LOMI	PCR	GSS	IC
1	FT	√					√				√		
2	CD	√	√	√	√			√			√		
3	CT										√		
4	CP			√									
5	MD								√	√			
6	MT							√					
7	MP			√									
8	MB/MGD	√		√				√				√	

influences the development or consequence of an accident in a significant way.

Since reactivity coefficients change during the life of the core, ranges of coefficients are employed in the analysis of transients, to determine the response of the plant throughout its life. The reactivity coefficients are calculated with qualified computer codes, approved analytical methods and calculation models. The effect of the 3D power distribution on core-average reactivity coefficients is implicit in these calculations.

The neutron balance can be upset by a perturbation to any one of the local parameters, such as neutron flux, fuel burnup, coolant density, fuel temperature, etc. In the analysis of hypothetical accidents, thermalhydraulic feedback must be considered, and the manner in which the shutdown systems act (separately) to terminate the power excursion must also therefore be carefully studied.

Because the uncontrolled generation of heat is the main concern in accidents, reactor-physics analysis is basic to safety analysis. As an essential input to downstream components of analysis, it provides calculated data on core state, power distribution, and neutronic trips for shutdown-system actuation.

Based on assumptions for the initiating event, physics analysts need to model/evaluate changes in the core state (nuclear properties, reactivity-device positions) to simulate the neutronic behaviour, i.e., calculate reactivity, the neutron-flux distribution in space and time, the bulk reactor power, the 3-D power distribution, the energy (heat) added to the fuel. In assessing shutdown-system performance, it is important to demonstrate that both SDS-1 and SDS-2, if actuated separately, can shut down the reactor (render and retain the core subcritical), and also show that the neutron flux and power go to zero everywhere, that energy addition to fuel does not lead to unacceptable consequences, and ultimately, that any radiological doses to workers and public are within regulator-prescribed limits.

9. Approach to Criticality

9.1 Source Neutrons

Source neutrons are essential for reactor restart after a long shutdown. The term "source neutrons" applied to a particular time interval refers to a steady supply of neutrons, constant over the time interval of interest. This supply must be independent of the current or very recent fission rate, which can vary over the time interval. Thus, source neutrons exclude prompt neutrons and even delayed neutrons which originate in the fuel (i.e., those born in the fuel itself). This exclusion does not apply to delayed photoneutrons, which come from fissions that have occurred a long time before, and whose numbers are quite constant over the current time interval (further discussion of this point below).

A CANDU therefore has two "built-in" neutron sources, one arising from the radioactive decay of the fuel itself, and the other from the presence of heavy water.

1. Spontaneous fission neutrons
   Neutrons from spontaneous fission come mainly from U-238, which constitutes most of the fuel. The contribution to reactor power from fissions induced by neutrons originating in spontaneous fission is constant, at about 10^{-12}% of full power. This low-level source is important only at first startup or after a prolonged shutdown, if no external neutron source is placed inside the reactor.

2. Delayed Photoneutrons
   The delayed-photoneutron source strength depends on the flux of energetic gamma rays that can break up deuterium nuclei, which in turn depends on how long the reactor has been operating. Following prolonged operation at significant power levels (>10% full power), the decay gamma flux is proportional to the power. The power from the delayed-photoneutron source is typically about 0.03% of the power level (compared with about 0.5% of the power

from the delayed neutrons). The delayed-photoneutron source becomes important following the disappearance of the delayed neutrons from the fuel after shutdown.

In this section, we are following the usual convention of treating delayed neutrons born in the fuel as "regular" delayed neutrons, distinguished from the delayed photoneutrons from photodisintegration of deuterium, which are labelled "source neutrons". The distinction rests on the practical operational effects that result from the different half-lives of the precursors of these kinds of neutrons.

The longest-lived precursor of "regular" delayed neutrons from the fuel has a 54-second half-life. Following a power change, it takes a few minutes for the delayed-neutron precursor bank to come to a new equilibrium. Precursors of delayed photoneutrons, in contrast, have much longer lifetimes. The longest half-life of delayed-photoneutron precursor group 1 is that of Ba-140, at about 12.8 days. Thus, the delayed-photoneutron source persists for several weeks after shutdown. Following a power change, it takes weeks for the delayed-photoneutron fraction to reach its new equilibrium strength.

For neutron balance in a critical steady-state core, it is convenient to lump the delayed photoneutrons with the delayed neutrons — they act like delayed neutrons with long lifetimes. Dynamic critical reactor behaviour depends on a delayed neutron "delay tank" — a neutron source that changes slowly (over many prompt-neutron cycles) following a reactivity insertion. The distinction between source neutrons and "regular" delayed neutrons depends on whether or not the neutron supply is nearly constant over the time interval considered. From an operational perspective, for the scenarios like power ramps that take minutes or hours, it is necessary to distinguish delayed photoneutrons from "regular" delayed neutrons.

9.2 Subcritical Multiplication Equation

In a critical reactor with no neutron sources other than induced fission, the neutron population in the reactor remains constant from one generation to the next since absorption and leakage balance exactly the neutron production. In a subcritical reactor without a neutron source, the neutron population would quickly collapse. However, when a neutron source independent of the flux level (also called an external source) is present, the fuel in the subcritical reactor acts as a multiplier of source neutrons, so that the actual power generated when a source is present is much greater than would be produced by the source neutrons alone. Source amplification in the subcritical reactor depends on its subcriticality, as shown below.

Assuming that a neutron source is inserting S_0 neutrons per unit time into a subcritical reactor whose effective multiplication factor, k_{eff}, is less than one, the total neutron population per unit time (S_∞) in the reactor will come to an equilibrium value:

$$S_\infty = 1/(1 - k_{eff}) \times S_0$$

With this subcritical multiplication, the neutron flux in the reactor will come to a steady state when there is an independent or external source in a subcritical reactor.

In fact, we can now say that there are only 2 situations in which there can be a steady state in a reactor:

- If there is no external or independent neutron source in the reactor, and the reactor is critical (reactivity $\rho = 0$), and
- If there is an external or independent source in the reactor, and the reactor is subcritical ($\rho < 0$).

With the multiplied neutron flux, there will be a higher power level in the reactor than that provided by the independent source on its own. It can be shown that the steady-state power level is given by

$$P_{observed} = \frac{P_{source}}{-\rho}$$

where

- P_{source} is the power that would be generated by the source neutrons themselves in the absence of any multiplication by the fission process (that is, if v, the number of neutrons released per fission, were equal to zero)

- $P_{observed}$ is the measured (actual) power level, and
- ρ is the reactivity ($= \rho = 1 - 1/k_{eff}$, see Section 3.1); here $\rho < 0$ and is a measure of the reactor's subcriticality.

The factor $1/(-\rho)$ is called the subcritical multiplication factor. The amount of subcritical multiplication depends only on the value of ρ (or of k_{eff}). For example, in a reactor that is well below critical, say $\rho = -0.100 = -100$ mk, the observed power level $P_{observed}$ is 10 ($= 1/0.100$) times greater than the actual photoneutron source power. If $\rho = -0.020 = -20$ mk, the subcritical multiplication factor will be 50 ($= 1/0.020$). Figure 19 shows the variation of the subcritical multiplication factor as ρ changes. When a source is present, most of the neutrons in the reactor (at least for $-\rho > 500$ mk) are not neutrons from the source, but neutrons that originate in fissions induced by source neutrons.

9.3 Application of the Subcritical Multiplication Equation

9.31. Dynamics in the Subcritical Core

A positive reactivity insertion in a subcritical reactor increases power to a new equilibrium level. The size of the increase and the time it takes for the power to stabilize at the new value depends on the reactivity

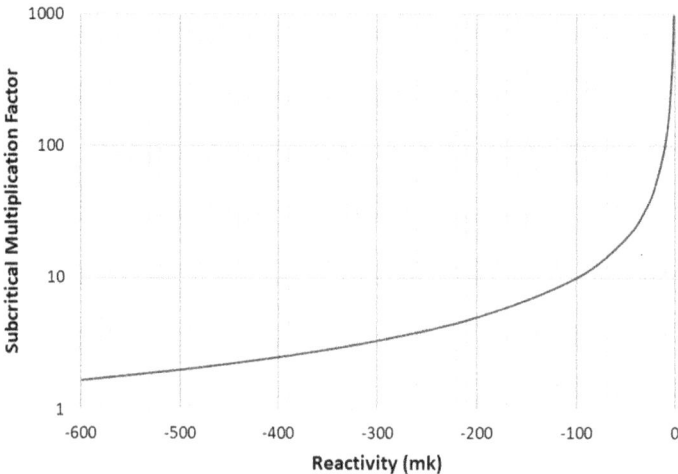

Figure 19 The Subcritical Multiplication Factor

inserted and on the final value of the negative reactivity. The smaller the reactivity, i.e., the closer ρ is to zero, the larger the increase in power for a given reactivity insertion, and the longer the time required to stabilize.

From the subcritical multiplication equation, it is straightforward to calculate that the subcritical power will be increased by 25% for the following two scenarios: a step increase in a deeply subcritical core from $\rho = -300$ mk to $\rho = -240$ mk, and a much smaller step increase in an almost critical reactor core from $\rho = -1.5$ mk to -1.2 mk. The 25% power increase takes less than 0.01 seconds for scenario 1 and about five minutes for scenario 2.

As the reactor approaches criticality, each successive reactivity insertion produces a larger power increase, and the time to stabilize gets longer. This produces a smooth transition between the subcritical and critical states. A deeply subcritical core is very sluggish; even large reactivity increases could go unnoticed. An almost critical reactor behaves a lot like a critical reactor.

9.3.2 Determining Core Reactivity and Reactivity Worth of a Device

The subcritical multiplication equation has some practical applications. One example is in using the subcritical multiplication equation to calculate the reactivity of the core by observing the power before and after a known reactivity change. Usually we do not know the source strength, but we do know it is constant in the short term. We can write the equation for each observed power level (or, alternatively, for each reading of a detector) and eliminate the unknown source strength to determine ρ. Once ρ is determined, the reactivity worth of a new device can be determined by observing the power level (or detector reading) before and after inserting the device.

9.3.3 Approach to Critical by Power Doubling

Another important practical application of the subcritical multiplication equation in reactor operation is the power-doubling rule, stated as "When a certain reactivity addition in a subcritical reactor causes reactor power (or detector count rate) to double, a further addition of the same amount of positive reactivity will make the reactor critical".

Reducing the subcriticality to half its value, i.e., replacing ρ with $\rho/2$ in the subcritical multiplication equation, causes the power to double. We can turn this statement around and say that when power doubling is observed, then the reactivity ρ has been halved to $\rho/2$. Obviously, then, adding more positive reactivity equal to $-\rho/2$ will make the reactor critical (ρ is reduced to zero and $k_{eff} = 1$). If the reactor is deeply subcritical (i.e. poisoned), the power doubling can be observed during poison removal (for example, by monitoring a detector's reading). From the amount of poison removed to cause a power doubling, the approximate poison concentration when the reactor will reach criticality can be estimated.

Table 8 demonstrates how a series of power doublings makes a deeply subcritical reactor approach criticality. The reactor is initially subcritical with ρ at -320 mk and measured power at 1.25×10^{-6} of full power. Ten successive doublings raise the power by three orders of magnitude

Table 8 Typical Series of Power Doublings
(*Source*: Table 3.3 from [4])

	$-\Delta k$ (mk)	power %F.P.	power (decades)
0	320	1.25×10^{-4}	$10^{-3.9}$
1st	160	2.5×10^{-4}	$10^{-3.6}$
2nd	80	5×10^{-4}	$10^{-3.3}$
3rd	40	1×10^{-3}	$10^{-3.0}$
4th	20	2×10^{-3}	$10^{-2.7}$
5th	10	4×10^{-3}	$10^{-2.4}$
6th	5	8×10^{-3}	$10^{-2.1}$
7th	2.5	1.6×10^{-2}	$10^{-1.8}$
8th	1.25	3.2×10^{-2}	$10^{-1.5}$
9th	0.6125	6.4×10^{-2}	$10^{-1.2}$
10th	0.31	0.128	$10^{-0.9}$

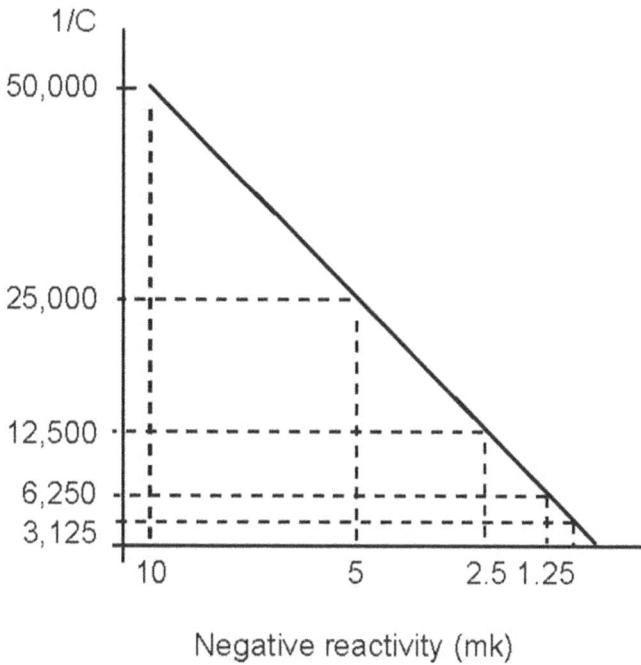

Figure 20   Approach to Critical by Power Doubling - Inverse Count Rate as a Function of Negative Reactivity (*Source*: Figure 3.7 from [4])

($2^{10} = 1024 \approx 10^3$). Each doubling leaves half the negative reactivity in the core, so the reactor is not yet critical (with ρ at -0.31 mk) after 10 successive doublings. When the reactor is only slightly subcritical (by 1 or 2 mk), doubling power can be handled within the available control range of the LZCs. If a small (≤ 10%) lowering of the liquid-zone level produces a power doubling, the operator knows the RRS could manoeuvre power to any demanded level at any demanded rate, and the reactor is therefore declared "critical".

Each time power doubles in the subcritical core, noting the change in reactivity that caused the doubling is, in effect, a measurement of ρ or k_{eff}. Figure 20 is a typical plot of inverse detector count rate (CR, proportional to the neutron power $P_{observed}$) monitored from the startup instruments for the initial approach to criticality of a subcritical reactor. The graph is a straight line, because 1/CR is proportional to ρ, based on the subcritical multiplication equation. The point where the extrapolated line crosses the axis represents criticality (when $\rho = 0$, $1/CR = 0$). The graph demonstrates visually that taking the reactor to criticality by power doubling (i.e., by cutting $1/CR$ in half), is a cautious way of approaching criticality.

As long as one doubles power, the reactor gets closer and closer to critical without actually going critical. By plotting the graph, or by keeping track of the reactivity changes that double power, the operator can predict the reactivity worth of the controlling device (such as LZCs for CANDU) at criticality.

10. Reactor Shutdown and Reactor Restart

10.1 Thermal Power, Neutron Power, Decay Power, and Fission Power

The power referred to most frequently in reactor physics is neutron power. Neutron power is essentially the fission rate multiplied by the average prompt energy released and recovered per fission (see Section 2.1.2). It is also called "prompt" power, as it appears very quickly following fission. We cannot measure neutron power directly, but we do monitor the neutron flux with ion chambers located outside the calandria and in-core flux detectors. These neutronic signals are calibrated to the thermal-power measurement which allows neutron power to be derived.

However, the actual output of the reactor is in the form of heat, and the heat transferred to the SGs is called reactor thermal power. Thermal power takes into account nuclear-decay heating and conventional heat (pump heat). Thermal power is not proportional to neutron power or flux level: the relationship between flux and thermal power is not linear, on account of decay heat, of heat generated by fluid friction (which is independent of flux level), and of heat lost to the moderator and shielding. Because of these non-linearities, periodic calibration of the neutronic signals against the thermal-power measurement is needed, to follow any power changes.

Thermal power has the advantage of being the actual, useful heat-power output of the reactor. Thermal power can be evaluated either on the reactor side (PHTS side) or on the SG side (Secondary-Heat-Transport side). On the rector side, thermal power is generally evaluated by measuring the PHTS flow rate and the temperature change from RIH to ROH. This measurement of thermal power has the disadvantage of an excessive time lag between neutron-power changes and detected thermal-power changes (around 25 s) and a non-linear relationship with neutron power, especially at low power levels.

Decay power (usually called decay heat), originates in the nuclear decay of fission products (with various half-lives ranging from fractions of a second to thousands of years) and appears with a delay (seconds/minutes to weeks/months) following fission. In steady state, decay heat is ~6-7% of the total thermal energy release and does not show up at the instant of fission. Consider that 6-7% of ~2,000 MW (or more) of heat is not negligible. Therefore, even after a reactor is shut down, energy (heat) from nuclear decay actually appears in the reactor for many hours, days, even months after the chain reaction is stopped; see the variation of decay heat with time in Figure 21. Decay heat is a concern in reactor and plant safety, because fuel cooling must still be provided after reactor shutdown or after fuel is removed from the reactor. If decay heat cannot be dissipated safely, the fuel and reactor core can seriously overheat, as happened at the Fukushima Daiichi accident in 2011.

Fission power is the name given to the heat (neutron power plus decay heat) generated in the fuel because of nuclear processes and hence it is proportional to the neutron flux when the reactor power remains constant. This includes heat deposited in the moderator and shielding. It does not include any conventional heating. The operating license

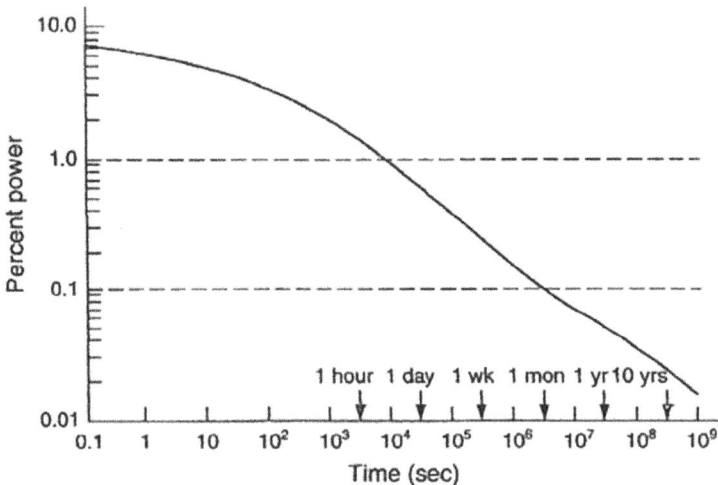

Figure 21   Decay Power vs. Time (*Source*: Figure 1-4 from [11])

places an upper limit on fission power, which is enforced by regulating the reactor thermal power.

Figure 22 illustrates the relationship among these powers for a CANDU 6. The numbers will differ from reactor to reactor, but the percent values are quite similar.

10.2   Reactor Power Rundown

When the reactor is shut down following extended high-power operation, even though the fission process stops more or less instantaneously, the thermal power of the reactor decreases much more slowly than the neutron power, mainly because of the decay heat. The thermal output immediately after shutdown is still ~6-7% of its full-power value; it will then decrease slowly as the fission products decay. Figure 23 is a graph of a typical rundown of neutron power and thermal power after a reactor shutdown (or trip). Although the fission rate falls off rapidly, decay heat can fall off only according to the decay rate of the fission products producing it.

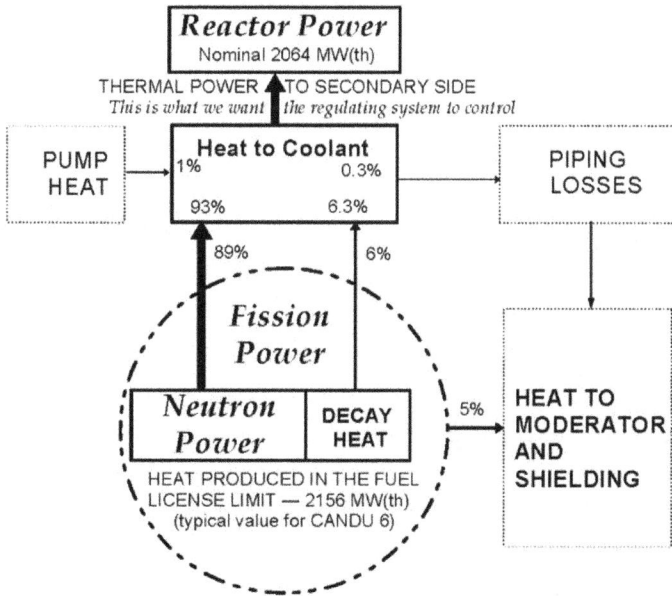

Figure 22   Typical Fission Power, Thermal Power, and Neutron Power for CANDU 6 (*Source*: Figure 7.1 from [4])

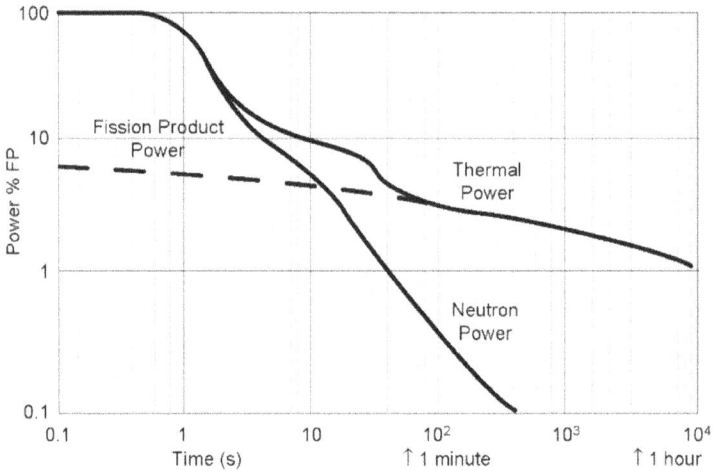

Figure 23   Change of Neutron Power and Thermal Power after Shutdown in CANDU

Typically, after a minute or so in a shutdown, the neutron power makes little contribution to the thermal power. Thermal power drops to about 3% full power in about three minutes (which explains, incidentally, why 3% is the capacity chosen for the auxiliary boiler feed-pump). It continues to drop to ~1.5% in an hour, and to below 1% over a period of ~8 hours. As a practical point, we should note another factor that slows the rate of decay of thermal power generated in the core. The heat-transport pumps generate heat at the rate of nearly 1% of full power. This heat source persists until it is possible to switch to the much smaller shutdown-cooling pumps.

As illustrated in Figure 23, thermal power and neutron power are not proportional when power is low. To protect the reactor against criticality accidents, neutron power must be monitored even at low power levels. Thermal-power measurement is incapable of protecting the reactor from a rapid increase of reactivity, in fact it is rather slow even for normal control.

We will now take a detailed look at the decay of neutron power after a shutdown. Figure 24 shows the time variation of neutron power following a shutdown. Here distinction is made between delayed neutrons

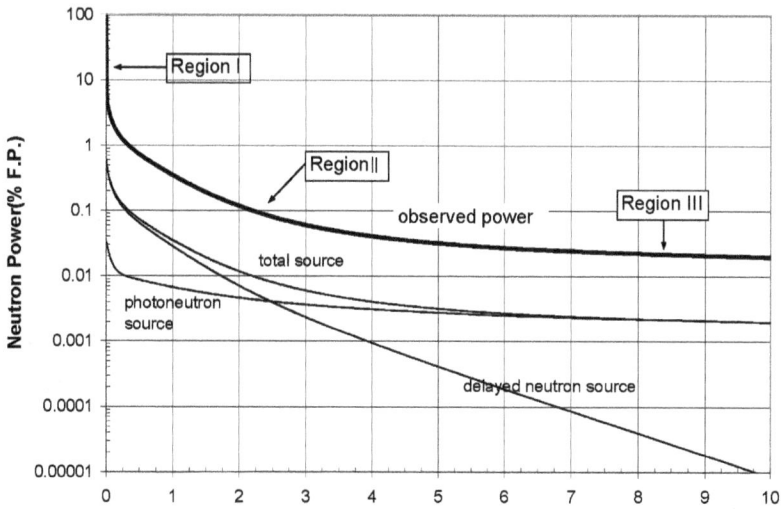

Figure 24 Variation of Neutron Power vs Time (in minutes) after Shutdown in CANDU (*Source*: Figure 7.4 from [4])

born in the fuel and delayed photoneutrons. The evolution of the neutron power can be divided into three distinct time intervals or regions as follows:

- Region I: Prompt collapse of prompt neutrons
    The fission rate falls off rapidly. More than 99% of neutrons are prompt and disappear very rapidly on shutdown, before the delayed precursors have had time to decay. The neutron population stabilizes temporarily at a level determined by the subcritical multiplication of the delayed neutrons in the reactor. The actual drop is always determined by the value of delayed-neutron fraction β and the amount of inserted reactivity. As the reactor is subcritical, the power drop appears throughout regions II and III.
- Region II: Delayed-neutron hold-up
    In region II the source of neutrons is the decay of the delayed-neutron precursors which were present prior to shutdown. The delayed-neutron-source power decreases rapidly initially as the shorter-lived precursors decay, and then decreases more slowly,

governed by the longest-lived precursor group, which has a half-life of ~54 s. The delayed-neutron-source power is negligible within 10 minutes after shutdown.

- Region III: Delayed-Photoneutron hold-up
  After shutdown, the delayed photoneutrons become the dominant neutron source in the reactor within 2 to 3 minutes. In a week or so, the only significant source comes from the group-1 delayed photoneutrons governed by Ba-140 which has a half-life of ~12.8 days. The delayed-photoneutron source keeps the reactor- power measurements on scale for several weeks after shutdown. Eventually (after a year or more) the delayed-photoneutron-source strength decreases to the point that it is comparable to the spontaneous-fission-source strength ($\approx 10^{-14}$ full power).

In practice, the transition between Regions II and III is arbitrary, as both the delayed neutrons and the delayed photoneutrons contribute to the source in Region II, but the balance between the two shifts steadily towards the delayed photoneutrons as time goes on.

10.3 Guaranteed Shutdown State (GSS)

A GSS is a state where the reactor remains in a stable, subcritical state independent of reactivity perturbations caused by any possible changes in core configuration, core properties, or process-system failures. Where possible, this shall be achieved without operator intervention. Possible reactivity perturbations which are taken into account are: temperature reactivity effects for hot and cold shutdown conditions; decay of the saturating fission products (Xe-135 and Pm-149, etc.); decay of Np-239 into Pu-239; draining of all LZCs; CVR under hot shutdown conditions; withdrawal of all adjuster rods; etc.

There are 3 types of GSS in the operating CANDU reactors: moderator-drained GSS, poison-based GSS, and rod-based GSS. Moderator-drained GSS, achieving GSS by dumping moderator from the calandria, was used only in earlier CANDU designs, such as Pickering-A, which has a moderator dump tank. Poison-based GSS, achieving

GSS by using moderator poison, is the traditional GSS widely used for planned maintenance outages. Rod-based GSS, achieving GSS by ensuring all absorber rods are inserted and locked in the core with a minimal amount of moderator poison as a double contingency measure, is a new GSS developed and used for unplanned and short-duration shutdowns only. The characteristics of poison-based GSS and rod-based GSS are summarized as follows:

Poison-based GSS

- Advantages:
  Bigger subcriticality margin: hundreds of mk of negative reactivity of moderator poison
  Used for planned maintenance outages
- Disadvantages:
  The solid absorber rods (SORs, MCAs, adjuster rods) are not credited
  Creation of more radioactive waste inventory
  Complex, time and labour intensive to prevent inadvertent dilution or precipitation of moderator poison:
    ✓ Purification off, sources of clean water isolated
    ✓ Continuous circulation
    ✓ pH and concentration checked regularly

Rod-based GSS

- Advantages:
  Safety benefits
    ✓ Reduction of doses to the workers (save up to 3 mSv from the outage collective dose)
    ✓ Environmental friendly - less chemical and radioactive waste
    ✓ Less material degradation (reduced thermal cycle on pressure tube and other reactor components)
    ✓ Easier and more robust GSS management
    ✓ Lesser Regulator's approval frequency

Economic benefit
- ✓ Reduced transition time (about one day shorter) from a GSS to full power (less poison removal)
- ✓ Reduced cost – saving ~Can$1M per outage

Disadvantages:
- ✓ smaller subcriticality margin, new GSS rods might be needed
- ✓ SDS2 is to be poised (available) all the time
- ✓ Extensive technical assessments required
- ✓ Prevention of inadvertent removal of solid rods or poison

10.4   The Shutdown State

During operation, a reactor is always critical or in a deeply subcritical GSS, or in active transition between these states. Once the reactor is cooled, placed in a GSS, and xenon has decayed (after 3 days or so), there are only small reactivity changes in the shutdown core. The observed power then trends downward with the gradual decrease in the source strength due to the decay of the delayed-photoneutron precursors. The observed power level in a subcritical reactor depends on the subcriticality and on the neutron-source strength, as per the subcritical multiplication equation (see Section 9.2).

After the reactor is placed in a GSS following extended high-power operation, temperature changes and the decay of fission products both affect core reactivity. Table 9 lists the effects and also gives typical CANDU values for each and the time over which the reactivity changes. The range of values in the table is intended to encompass the real variations from reactor to reactor and uncertainties in some of the parameters. The net effect of these changes is that when the reactor returns to the hot shutdown state, ready for restart, the core reactivity discounting the poison is +30 to +32 mk higher than it was before shutdown. Moderator poison keeps the reactor deeply subcritical. On restart, excess poison is removed and the poison concentration at criticality is exactly that needed to offset the increase in reactivity that has occurred during shutdown.

Table 9 Typical Reactivity Changes Following Shutdown for CANDU (*Source:* Table 7.2 from [4])

	Reactivity Effect (mk)	Time Scale
Temperature Effects		
Full Power Hot to Hot Shutdown	+3	seconds to minutes
Hot Shutdown to Cold Shutdown	-4 to -6	minutes to hours
Fission Product or Precursor Decay		
Xenon Decay	+28	80$^+$ hours
Np-239 Decay	+9 to +12	2 weeks
Pm-149 Decay	-8 to -11	2 weeks
Other Saturating FP Decay	-0.5 to -1	5 to 10 days

Off-line calculations to predict these changes in reactivity are unlikely to be more accurate than ± 0.3 mk (1% of 30 mk). In practice, predictions of critical poison concentration are often no better than ± 1 mk, equivalent to ± 15% liquid-zone level or ±0.04 ppm Gd. The problem of accurate poison concentration measurements compounds the uncertainty. The reason for mentioning this is to point out that it is not possible to know the exact poison concentration at criticality in advance. Caution has to be in place to approach criticality as described below.

10.5 Restart from a GSS

Any restart from a GSS requires manual operation of the purification system to remove moderator poison. The monitoring instruments differ depending on the length of shutdown. For restarts within a few weeks of shutdown, the RRS instruments monitor the startup; for restarts after an extended outage, special startup instruments monitor the initial stages of startup, with a transfer to the RRS instruments after they come on scale. Reactor-power-monitoring instruments go off scale some time after shutdown. This time, typically a few weeks, depends on the particular reactor, its operating history, and the depth of its GSS.

10.5.1 Restart Within a Few Weeks of Shutdown – RRS is On Scale

With the normal RRS ion chambers in range, the RRS responds, but the reactor is not sufficiently close to critical to say that the RRS is in control. The effect of valving-in purification with the RRS holding power is to raise the liquid-zone levels with the net reactivity of the core unchanged. With purification valved out the poison concentration does not change, and a request to raise power decreases the liquid-zone level, and reactivity increases. Procedures differ in detail from station to station, but usually these two processes alternate. This avoids the complication of having simultaneous device operations to change core reactivity.

These two steps, the RRS response to a request to double the present power with purification stopped, followed by poison removal with the RRS holding power, are repeated until RRS achieves a power doubling when the operator requests it. The process is then iterated two (or three) more times, achieving three (or four) definite power doublings in all.

As explained in Section 9.3.3, one of the nicest features of this procedure is that the reactor cannot actually go critical! It is a procedure for approaching criticality, cautiously, while ensuring that the reactor necessarily remains subcritical. As long as the reactor is subcritical and the operator requests a power doubling, the RRS adds only half of the reactivity required for criticality.

10.5.2 Restart after an Extended Outage – RRS is Off Scale

During extended shutdowns, where the power drops to a very low level, the normal instrument readings are unreliable because background gamma rays contribute to the readings. The installed instruments are considered "off-scale" somewhere between about 10^{-6} and 10^{-7} of full power. The readings below this level are not proportional to the flux. Supplementary "startup instruments" are installed before this happens, and the initial stages in the subsequent approach to critical would use these counters.

Here, the initial approach to criticality following removal of the GSS uses the purification system to increase reactivity and the startup instruments to monitor the count rate. As there is no automatic regulation, the operating staff must perform the functions of the automatic system. This means observing power level, making sure that changes in power are within the expected range, and adjusting purification flow appropriately. Monitoring usually includes plotting a graph of inverse detector count rate vs. reactivity (or moderator-poison concentration), as described in Section 9.3.3. At some point in this process, the normal RRS instruments come on scale. The operator then uses the RRS ion chambers, the LZC system, and the purification system to take the reactor critical as described in Section 10.5.1.

10.6   Restart from a Non-GSS

There are two situations in which the reactor is not in a GSS before restart. These are poison override (restart within ~30 minutes or so of a trip) and last stages coming out of a poison outage (typically 35 to 40 hours after a trip). In each case, the restart uses the RRS, but the high, rapidly-changing xenon concentration adds complications (see Section 7).

10.6.1 Restart by Poison Override

Immediate recovery following a trip is possible, if at all, only after an SDS1 trip. For SDS2 this will take nearly as long as the poison-out time on account of the slowness of the poison removal. Assuming restart is an option following an SDS1 trip, the SORs must first be withdrawn to repoise SDS1. The reactor stays subcritical because of the buildup of xenon and because the liquid zones fill and MCAs drop into the core on a trip. The MCAs must also be withdrawn (in banks), followed by removal of adjuster rods, one bank at a time, until there is sufficient reactivity to overcome the xenon buildup. Suppose criticality is reached (by a decrease in the liquid-zone levels) following removal of the last bank of adjuster rods. A request to increase power begins the xenon burnout process. The liquid-zone levels rise again as xenon load decreases, and the RRS will request adjuster in-drive (one bank) each time the average liquid-zone level reaches 80%.

10.6.2 Restart after a Poison Outage

In this scenario, we assume that there are sufficient source neutrons to keep the RRS instruments on scale, and the RRS is available to control reactor power. During a poison outage, reactivity changes due to the decay of xenon are not under direct control. The reactivity changes are affected mainly by the xenon transient, whose characteristics are highly dependent on the operating history of the reactor prior to shutdown. When the xenon concentration starts decreasing, the reactivity starts to increase, and the ion-chamber signals increase. The RRS is in control of bulk reactor power throughout, since the instruments are on scale. As more xenon decay occurs, the liquid zones will start to fill to maintain the reactivity balance. Once the liquid zones reach their control limit, and the xenon decay continues, reactor power would increase in the absence of further control action. Poison addition and/or adjuster reinsertion (if they are out of core) will be required to maintain the liquid zones in control range.

11. CANDU Reactor-Physics Analysis Methods and Computer Codes

11.1 Evaluation and Processing of Basic Nuclear Data

Reactor physics aims to understand accurately the reactivity and the distribution of all the reaction rates (most importantly of the power), and their rate of change in time, for any reactor configuration. To do this, the multiplication factor (or, equivalently, reactivity) and the neutron-flux distribution under various operating conditions and at different times need to be calculated repeatedly. Most of the other parameters of interest (such as neutron reaction rates, power, heat deposition, etc.) are derived from them. They are governed by the geometry, the material composition and the nuclear data (i.e., the neutron cross sections, their energy dependence, the energy spectra and the angular distributions of secondary particles, etc.). For radiation-shielding calculations, additional photon interactions and coupled neutron-photon interaction data are required.

The problems related to the determination and verification of nuclear data used in reactor calculations concern specialists in nuclear physics rather than reactor physics. That is why we do not intend to expand on this topic here. It is important, however, for a reactor physicist to have a basic knowledge of this subject. To determine cross sections and other nuclear data is not simply a measurement problem; the measurement results need to be evaluated (selected and/or weighted), and any missing information must be "filled in" using nuclear models, the data must be placed in a standard format and processed for the purpose of use in reactor-physics calculations and, finally, the calculations must be qualified against neutron-physics experiments. These experiments are called integral experiments to distinguish them from differential experiments, and they can be a valuable source of additional information beyond that provided by direct nuclear measurements.

There is now a huge body of data collected from nuclear measurements performed by specialists over the past 8 decades, on various nuclides, for different reactions and incident-neutron energy. Because of the large amount of data and the necessity to choose between redundant measurements (or to average them using appropriate weighting) and to fill in any gaps, it was necessary to organize this information and to standardize the way it is presented and the procedures for its use. Evaluation bodies are responsible for this. The three main evaluations currently used by reactor physicists are listed below:

- ENDF/B (Evaluated Nuclear Data File), USA;
- JEF (Joint European File), Europe;
- JENDL (Japanese Evaluated Nuclear Data Library), Japan;

In these evaluations, nuclear data are presented with all the available known detail. This presentation is not necessarily the most appropriate for reactor-physics codes. For example, these codes often do not operate using point data (continuous energy curves), but instead use multiple-group data (curves that are averaged over energy ranges). Likewise, the rather complicated processing of the resonances of heavy nuclei generally involves pretabulation. Accounting must also be done for Doppler broadening and neutron thermalization. Special utility software must be developed to handle all of these aspects. This software provides an interface between the files supplied by evaluators and the actual reactor-physics code.

11.2   The Neutron-Transport Equation and its Solution Methods

To be able to do all the calculations needed in reactor physics, we need to have knowledge of the most basic flux quantity, which is the angular neutron flux, i.e., the flux defined for all locations, all directions of motion, and all energies. The equation for the angular flux is the neutron-transport equation (also called Boltzmann equation), which has seven independent variables:

- three spatial variables, such as x, y and z, to identify the position of particles;
- two directional variables, such as θ and φ, to identify the direction of motion of the particles;

- one energy variable, such as E, to identify the energy (or speed) of the particles; and
- one time variable, t, to specify the instant at which the observation or calculation is made.

On account of the complex geometry and structure prevalent in practical problems, as well as the complex variation with neutron energy of the nuclear cross sections of the various materials considered, it is difficult, or even practically impossible, to find a detailed numerical solution of the neutron-transport equation, even with a very fast computer. The detailed formalism of the neutron-transport equation and the solution methods are out of scope and are not presented in this monograph. Interested readers are encouraged to read related books. The purpose of briefly introducing the neutron-transport equation here is simply to make the reader aware of its complexity and to highlight the numerical methods and the corresponding computer programs (codes) that have been used for the reactor-physics design, operation, and safety analysis of current CANDU reactors.

The neutron-transport equation can be written in integrodifferential form (integrals on velocity (or energy) and direction, and derivatives in space and time) or in fully integral form. These two forms are strictly equivalent from a mathematical point of view, which means that the theoretical solutions are the same. In practice, however, there are differences: firstly, because we often make approximations, and not necessarily the same ones in the two approaches, and secondly because the solutions are rarely analytical, and so we must settle for numerical processing, which obviously takes different forms according to the operator to be processed.

As with other physical or engineering problems, the methods for solving neutron-transport problems can be classified into two categories: deterministic methods and stochastic (Monte Carlo) methods.

In the deterministic method, a mathematical model, established according to the physical properties of the problem, can be represented by one equation or a set of determined mathematical equations. To solve the neutron-transport equation (in either form) with the deterministic method, the seven independent variables are discretized, and approximate methods are used to find an exact or approximate solution. Table 10 lists the approximate methods commonly used in treating these independent variables. Figure 25 summarizes numerical methods used in solving the neutron-transport equation. Obviously, these methods can be reasonably selected and matched according to the specific situation of the problem and the requirements of the calculation. For example, RFSP and SORO solve the two-group neutron-diffusion equation (which is a simplification of the neutron-transport equation) over the whole core with a finite-difference method, while WIMS-AECL and DRAGON solve the multigroup neutron-transport equation at the lattice-cell level with the collision-probability method.

Table 10   Approximate Methods Used in Treating Independent Variables of the Neutron-Transport Equation (*Source*: Table 1.2 from [29])

Independent variable	E	r	Ω
Approximate method	• Few groups • Multiple groups • Ultra-fine groups	• Finite-difference method • Nodal method • Finite-element method	For the integrodifferential equation • Spherical harmonic approximation (P_N) • Diffusion approximation • SP_N approximation • Discrete coordinate method (S_N) • Method of characteristics For the integral equation • Collision-probability method • Interface-current method

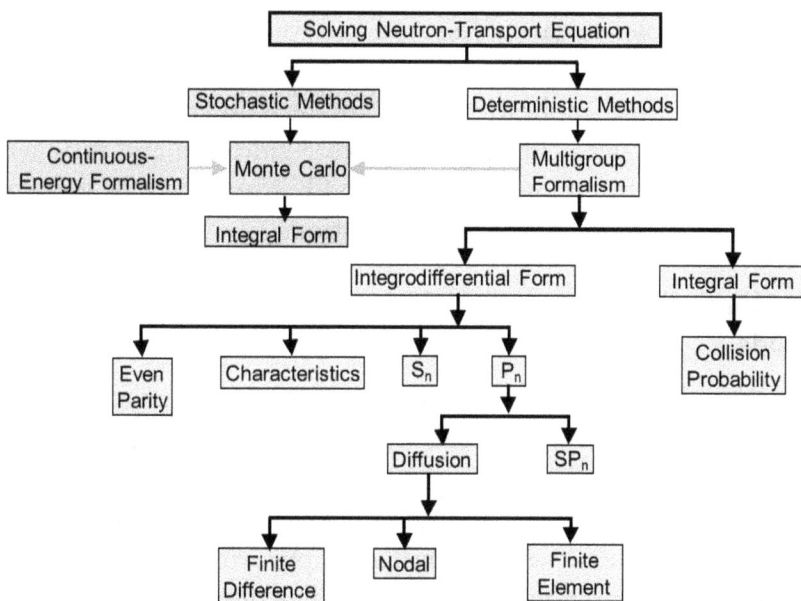

Figure 25   Summary of Numerical Methods Used in Solving the Neutron-Transport Equation

The stochastic (Monte Carlo) method is not a deterministic method, but rather a numerical method based on statistics or probability theory. The principle of the Monte Carlo method is to simulate neutron "histories" or "fates" as accurately as possible by randomly selecting the outcome of events as real neutrons do, i.e., according to the laws of nuclear physics: emission point, collision points, type of atom impacted, type of reaction, exit configuration from an event, etc. This approach remains statistical since the greatest possible number of neutron histories will be simulated in order to obtain the rates in which we are interested, with the greatest possible statistical accuracy. The advantage of the Monte Carlo method is that it does not necessarily require simplifications to the geometry, composition, and distribution of nuclei. In this sense, it can be described as "exact". In practice, however, it is only a reference method at best, because of uncertainties in the data as well as statistical uncertainties, which can never be reduced to zero. To reduce statistical uncertainties requires huge computer time and memory, which are major obstacles to the widespread and routine use of Monte Carlo methods. Therefore, the Monte Carlo method is usually used for occasional, non-routine applications, and for benchmarking. However, in some reactor-physics fields, such as reactor-shielding and fusion-reactor calculations, the Monte Carlo method has proven to be very useful and has been adopted for routine calculations.

11.3 Equivalence Approximations Adopted in the Deterministic Methods

We would ideally like to perform deterministic calculations on the actual, complex problem handled with the exact theory; but this cannot be attained at a reasonable calculation cost. Thus, the calculation is replaced with a simpler theory and a simplified problem (but still close to the actual problem), with equivalence approximations. By choosing the values of a certain number of numerical parameters carefully, we can let the simplified model preserve the results we deem to be essential of the exact model. This is the equivalence.

The concept of equivalence is in fact very general and is essential in making approximate solutions of the neutron-transport equation with deterministic methods, regardless of the type of reactor. Though it is not straightforward to estimate the residual error of the "equivalent" model, it is important to be aware of the following examples of equivalence approximations [12] adopted implicitly and explicitly in deterministic methods, before we introduce these methods for CANDU:

- Equivalence of multigroup energy condensation to continuous energy;
- Equivalence of few-group (such as 2-group) energy condensation to multigroup;
- Equivalence of homogenized fuel lattice to heterogeneous fuel lattice;
- Equivalence of 6 groups of delayed neutrons and 11 groups of delayed photoneutrons to about a hundred processes;
- Equivalence of the multiplication factor and lifetime of the point kinetics model to spatial kinetics;

- Equivalence of the simplified burnup chain to the real burnup chain;
- Equivalence of a defined "pseudo fission product" to the multitude of fission products neglected in the calculations;
- Equivalence of two effective fuel temperatures to the Doppler-effect calculation;
- Equivalence of the neutron-diffusion equation to the neutron-transport equation via Fick's Law;
- Equivalence between the isotropic treatment and a linearly anisotropic treatment via the transport correction;
- Equivalence of an infinite homogeneous reflector to the real reflector;
- etc.

11.4   Deterministic Methods for CANDU

Though direct solution by deterministic methods of the 3D multigroup neutron-transport equation with full complex geometry is feasible, this approach has not yet been used routinely for commercial reactors. Owing to the complexity of a power reactor, the current deterministic methods for full-core physics analysis of LWRs usually rely on a three-step approach that consists of

1. Generation of a multigroup nuclear-data library,
2. Two-dimensional (2D) physics treatment (neutron-transport equation) at the fuel-lattice level (see Figure 26), and
3. Continuum-physics approach (neutron-diffusion equation) at the 3D full-core level (see Figure 26).

Unlike this LWR approach, the physics analysis of a CANDU reactor is a four-step approach. An additional step is necessary to determine incremental changes of homogenized cross sections in the vicinity of reactivity devices, which are perpendicular to the fuel channels (see Figure 27). At each step, the calculations are carried out by dedicated computer codes that account for the characteristic features of CANDU reactors, such as cluster-type fuel, large volume of heavy-water moderator, 3D arrangement of reactivity

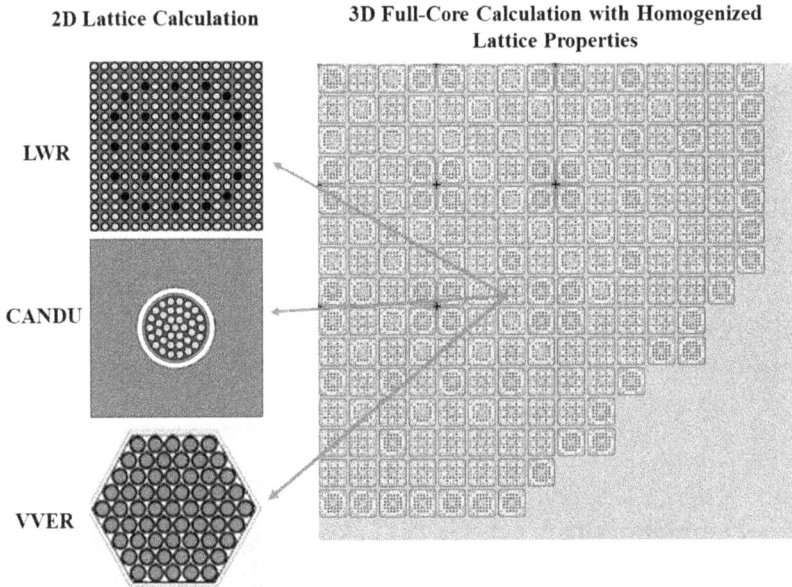

Figure 26   Sequential 2D Lattice and 3D Full-Core Calculations with Deterministic Methods

devices, and on-power refuelling in channels with bidirectional coolant flow and bidirectional refuelling.

Deterministic reactor-physics analysis begins with multigroup, microscopic cross-section libraries as the first step. These libraries have usually been processed from extensive tabulations (e.g., ENDF/B) or from reactor measurements. The cross sections in such libraries may have been subsequently modified from ENDF/B data in order to improve agreement between measurements and results computed by downstream steps.

The second step consists of lattice calculations for the bare lattice, i.e., for basic unit cells containing fuel, coolant, pressure and calandria tubes and the surrounding moderator volume, but excluding the representation of any interstitial reactivity devices, as shown in Figure 28. The lattice calculation is a collection of numerical algorithms and models which are capable of representing the neutronic behavior of a single lattice cell (or sometimes of multiple lattice cells) in a nuclear reactor in

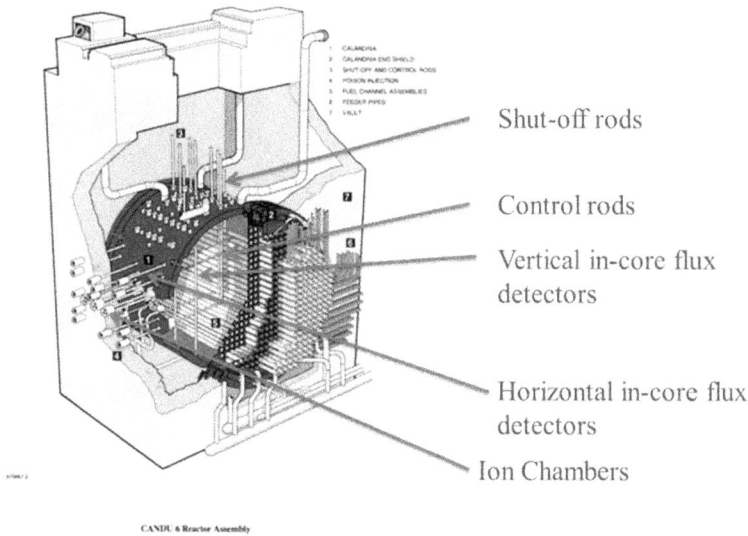

1 CALANDRIA
2 CALANDRIA END SHIELD
3 SHUT-OFF AND CONTROL RODS
4 POISON INJECTION
5 FUEL CHANNEL ASSEMBLIES
6 FEEDER PIPES
7 VAULT

Shut-off rods

Control rods

Vertical in-core flux detectors

Horizontal in-core flux detectors

Ion Chambers

CANDU 6 Reactor Assembly

Figure 27 CANDU Reactor Core

2D (or sometimes 3D). It is usually carried out in a detailed geometrical model, with a detailed representation of the neutron spectrum, and for all relevant local conditions, such as fuel temperature, coolant density etc. In performing lattice calculations it is normally necessary to repeat the lattice spectrum calculation at different burnups in order to account for the effect of changes in nuclide inventory. The lattice calculation provides homogenized properties for the lattice cell (or for each lattice cell in a multiple-cell configuration) in a few (usually two) energy groups, for input to the finite-core calculation.

The third step consists of supercell calculations in 3D, to determine the effect of interstitial reactivity devices on homogenized properties of cells which contain the devices. This effect is cast in the form of "incremental" cross sections in a few (usually two) energy groups, which are to be added to the bare-lattice cross sections (calculated in the second step) of lattice cells traversed by a reactivity device. In the CANDU world the term "supercell calculation" generally refers to a 3D transport calculation performed on a model that involves two lattice cells containing fuel bundles running horizontally and a reactivity device located

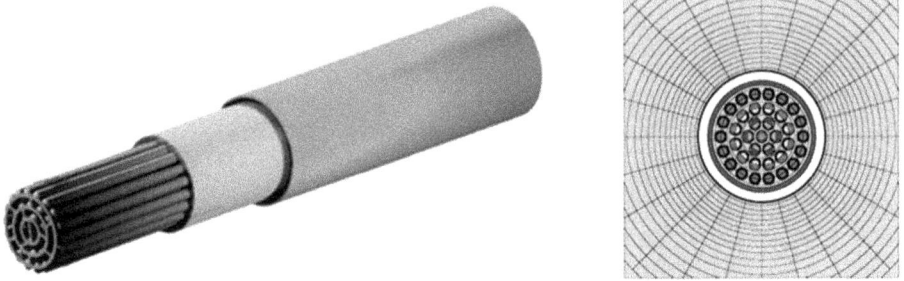

Figure 28 WIMS-AECL 2D Lattice Model for the CANDU Fuel Lattice

midway between the two fuel clusters (see Figure 29). Once the transport solution has been obtained, it is used to homogenize and condense the cross sections over the zone of interest in the presence of the reactivity device. The device incremental cross sections are determined by performing two supercell calculations, one with the device included in the model, and another with the device excluded, and then subtracting the homogenized cross sections obtained in the two cases. This information is then used in the finite-core calculation to locally correct the cross sections associated with cells affected by the reactivity devices.

The fourth step is the 3D neutron-diffusion calculation for the finite core, as shown in Figure 30. This calculation (in 2 or more energy groups) is based on a simplified representation of the reactor by a set of homogeneous rectangular cells, using lattice-homogenized cross sections obtained in the second step and device incremental cross sections obtained in the third step. The neutron-diffusion code calculates the finite-core multiplication constant and the global distribution of flux and power.

11.5 CANDU Reactor-Physics Deterministic Codes and Libraries

Significant time and effort have been invested by the Canadian nuclear industry to develop and maintain the reactor-physics deterministic codes that are used for CANDU analysis [13-14]. In particular, the development of the 2-D lattice code WIMS-AECL, the 3-D reactivity-device code DRAGON, and the 3-D finite-core code RFSP has taken much

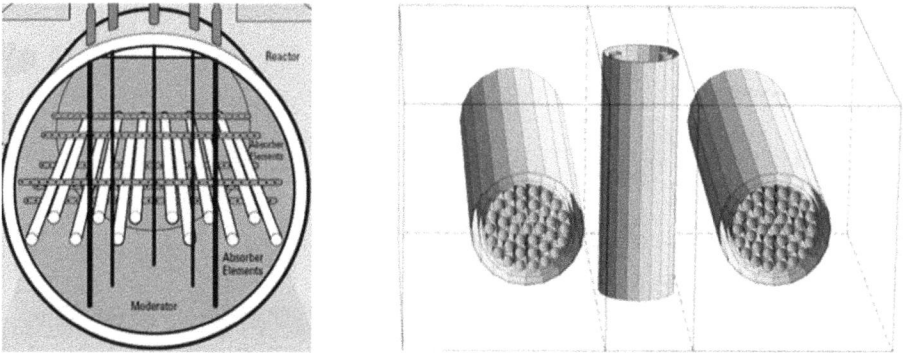

Figure 29 DRAGON 3D Supercell Model for the CANDU Reactivity Device

Figure 30 RFSP CANDU-6 Full-Core Model with Homogenized Lattice Properties

effort. Maintenance of the reference data libraries has also received significant attention by the industry in the past decades. These 3 Industry Standard Toolset (IST) codes, with associated libraries and interfacing codes, are used extensively for CANDU deterministic reactor-physics simulations, as shown on the left side in Figure 31. By experience, the WIMS-AECL/DRAGON/RFSP computational system has been found to give generally very good results. This suite of deterministic codes represents the state-of-the-art in CANDU analysis and addresses the regulatory expectations in terms of adoption of modern codes and methods.

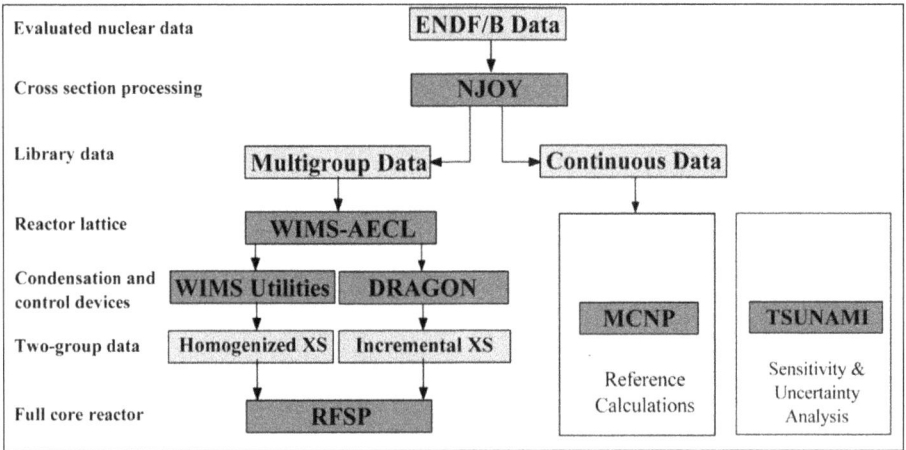

Figure 31 Neutronic Analysis Methodology for CANDU (Source: Figure 1 from [26])

Note that Ontario's CANDU utilities currently generally use the neutron-diffusion code SORO instead of RFSP for CANDU fuel-management calculations.

11.5.1 ENDF/B Nuclear-Data Library

Nuclear data is a fundamental component of any reactor-physics analysis. In the CANDU industry, lattice calculations are based on nuclear data originally described in the ENDF/B data set compiled by the Cross Section Evaluation Working Group (CSEWG). These data sets are then fed to a cross-section-processing code, such as NJOY, to produce the multigroup microscopic cross-section library to be used by the lattice codes.

The ENDF/B data sets are the building blocks for all of the reactor physics, shielding and criticality-safety analyses for nuclear reactors and facilities. The ENDF/B library is periodically updated to reflect results from new cross-section measurements, evaluations and validation exercises, and to address any deficiencies found in the previous library release.

The ENDF/B-VI nuclear-data library was released in 1990 and revised in eight interim releases. WIMS-AECL 2.5d with the

ENDF/B-VI-based nuclear-data library has been validated and applied for all domestic CANDU reactor-physics calculations. In December 2006, CSEWG released the ENDF/B-VII.0 nuclear-data library. WIMS-AECL 3.1.2 with the ENDF/B-VII.0-based nuclear-data library has been validated and is currently being applied for all domestic CANDU reactor-physics calculations. The ENDF/B-VIII.0 nuclear-data library was released in February 2018. The ENDF/B-VIII.0 data library benefits from recent experimental data and improvements in theory and simulation with notable advances for isotopes important to CANDU applications. The evaluation of the performance of the ENDF/B-VIII.0 nuclear-data library was initiated in the CANDU industry in 2018.

11.5.2 WIMS-AECL

WIMS-AECL[15-16] is a 2D multigroup collision-probability neutron-transport code routinely used for lattice-cell calculations of CANDU-type reactors. Owing to both the method's generality and its input flexibility, it can be applied to the numerical solution of a variety of reactor-physics problems. The code is particularly efficient for calculations in channel-type reactors with cluster fuel, as is the case with the CANDU lattice. The main features of the lattice-cell code WIMS-AECL are:

- Library: 89-group libraries based on ENDF/B versions V, VI and VII
- Geometry: 2-D, single-cell with fuel cluster, multiple cells with fuel cluster, off-center fuel-cluster cell
- Method of Solution: 2-D collision-probability method in the library or condensed energy-group structure
- Advanced Resonance Treatment: distributed self-shielding effects
- Other Features: visualization of 2-D geometries

11.5.3 DRAGON

DRAGON [17-18] is a 2D/3D multigroup collision-probability neutron-transport code developed at Ecole Polytechnique de Montréal. The main features of the supercell code DRAGON are:

- Library: Multigroup libraries based on ENDF/B versions V, VI, and VII

- Geometry: Capabilities to define complicated 2-D and selected 3-D cluster geometries, multiple cells with cluster structures
- Method of Solution: 2-D/3-D collision-probability method, method of characteristics, along with other methods
- Resonance Treatment: Advanced self-shielding capabilities including sub-group method
- Other Features: visualization of 2-D geometries

11.5.4 RFSP

RFSP [19-20] is the main scientific code for full-core neutronics simulation and analysis of CANDU reactors. The main function of RFSP is to calculate both static and kinetic neutron flux and power distributions in the core, using two-energy-group neutron-diffusion theory in 3D. It has a wide variety of functionalities: time-average simulation for reactor design, time-dependent refuelling simulation, slow transients (for xenon), fast kinetics calculations (such as for LOCA), control and shutdown system modelling, calculation of harmonic modes, flux detector responses and flux mapping, etc. The main features of the core-analysis code RFSP are:

- Problem Solved: Two-group time-independent and time-dependent neutron-diffusion equations for the eigenvalue (the inverse of the multiplication constant) and fixed-source problems, flux-mapping equations
- Geometry: 3-D, Cartesian geometry
- Method of Solution: Finite-difference method for the neutron-diffusion equations
- Cross-Section Model: History-based local-parameter method, macro- and micro-depletion methods, multicell methodology
- Other Features: Coupled to CANDU thermalhydraulics codes, unique capabilities for the design and analysis of CANDU

11.5.5 Coupling WIMS-AECL and RFSP via Various Cross-Section Models

RFSP requires the few-energy-group lattice cross sections calculated with WIMS-AECL in the form of fuel tables. However, WIMS-AECL does not produce fuel tables in a format that can be used directly by RFSP. WIMS Utilities is a collection of Fortran programs and Perl scripts, used to postprocess WIMS-AECL results and generate cross-section tables for use in RFSP.

For an accurate calculation of the power distribution in a CANDU-type reactor, it is important to generate few-group homogenized cross sections for each fuel bundle in the core, accounting for local conditions (fuel temperature, coolant density, etc.) and the burnup history of the fuel bundle, as well as the effects of the environment if necessary. The cross-section models used in RFSP for CANDU core analysis have evolved starting from the uniform-parameter method, through the grid-based local-parameter method, up to the history-based local-parameter method.

The uniform-parameter method is the simplest technique, in which the lattice-cell properties are functions of burnup only, by assuming effective core-average conditions for all fuel bundles or fuel assemblies. The grid-based local-parameter method (also called the macro-depletion method in LWR applications) [21] is an improved method in which the lattice properties are functions of the fuel burnup and other local parameters. Since the macroscopic cross sections used in the interpolation are generated beforehand at various assumed operating conditions, the grid-based local-parameter method does not take into account the dependence of the cross sections on the fuel bundle's or fuel assembly's history.

The history-based local-parameter method (also called the embedded lattice model in the recent literature) [22], developed in RFSP for simulating the CANDU reactor cores, is one of the most advanced cross-section models used in production codes. With this method, the lattice-cell code is coupled directly with the core-analysis diffusion code, and the lattice-cell calculations are performed for each fuel bundle or fuel assembly at each time step, so as to treat local parameters and the history of each fuel bundle individually. This method relies, in practice, on the premise that the nuclear cross sections associated with a given reactor state can be calculated quickly with sufficient accuracy using a simplified lattice-cell code such as POWDERPUFS-V used previously or Simple-Cell Model (SCM) [23] currently used. It is still currently impractical for routine history-based calculations to use nuclear properties from a modern lattice-cell code such as WIMS-AECL directly because of the computational effort required. The SCM history-based method proved adequate for the calculation of the CANDU reactor with the natural-uranium fuel. However it shows significant errors when used for other types of fuel, such as the Advanced CANDU Reactor (ACR) fuel, which are significantly different from the CANDU natural-uranium fuel.

In order to use the WIMS-AECL results directly without appealing to the SCM surrogate, the micro-depletion method [24] was developed and implemented in RFSP. It is an improved method compared to the traditional macro-depletion method because it tracks, at the core analysis level, both the microscopic cross section of a nuclide and its number density, which depend on the depletion history. The micro-depletion method is state-of-the-art and proven technology that is used widely throughout the nuclear industry. This method was implemented in RFSP and verified by comparison with WIMS-AECL results for the CANDU natural-uranium fuel and the ACR fuel.

The homogenized lattice properties employed in RFSP are usually calculated for a single lattice cell (considered as a heterogeneous medium) with reflective or periodic boundary conditions, without considering the effect of the environment due to nearby lattice cells. In some cases, multiple lattice cells (multicells) are needed to correct for the effect of the neighbourhood on a single lattice cell or the effect of the presence of the reflector for advanced CANDU reactor or CANDU reactor with advanced fuel cycles. This is especially important in configurations where there are large heterogeneities in the reactor core, such as the core-reflector interface and checkerboard-lattice structure. With the development of the multicell capabilities in WIMS AECL version 3.1, the multicell methodology was developed in RFSP-IST to account for the effects of the lattice-cell environment for 3D static and kinetics simulations, while maintaining the basic structure of the conventional single-lattice-cell-based reactor-physics homogenization used for CANDU reactor calculations for decades [25-26].

11.6 Monte Carlo Methods and Codes for CANDU

11.6.1 Advantages of the Monte Carlo Method

The deterministic method is not the only one for doing reactor-physics calculations. The Monte Carlo (stochastic) method can also be used to solve the neutron-transport equation in the fully integral form. The advantages of Monte Carlo methods are numerous:

- The method is suited to solve complicated 3-D neutron-transport problems by a single-level process with fewer approximations
- It has the capability of directly simulating the physics process with fewer limitations to the geometry
- The rate of convergence is independent of the problem dimension
- The method, by nature, is ideal for computer parallel processing
- Almost all relevant physics (with neutrons, photons, electrons, etc.) can be included.

11.6.2 CANDU Reactor-Physics Monte Carlo Codes

On account of significant advances in the computing performance of hardware and software in the past decades, there is growing demand for using Monte-Carlo codes to model and simulate CANDU full-core problems. Once reliable Monte-Carlo CANDU full-core results are computed, they can be used to evaluate the accuracy of the deterministic codes, especially for the calculation of key parameters that are hard to measure in operating CANDU reactors, such as CVR, channel-power and bundle-power distributions, etc.

It should be noted that, even though significant efforts and progress have been made in the past, many issues still remain in Monte-Carlo calculations of very large systems such as CANDU full-core problems - issue such as huge computing resources, large memory requirement, slow fission-source convergence, fission-source correlation, etc. due to their higher dominance ratios[2]. These issues were not observed in the problems with a smaller dominance ratio such as a fuel assembly or critical facilities. These issues are more challenging for the prediction of tally results (such as channel-power and bundle-power distributions) than the prediction of integral results, such as the k-effective values.

MCNP5 (Monte Carlo N-Particle, version 5) [27], developed by Los Alamos National Laboratory in the USA, has been qualified as an IST code for CANDU applications. The accuracy of CANDU deterministic calculations may be checked by comparison with measurements or against results of MCNP, as shown on the right side in Figure 31. In the past decades, MCNP has been extensively used for some specific (time-independent) CANDU and Advanced CANDU Reactor (ACR) calculations of reactivity, reactivity coefficients, channel/bundle power distribution, in-core device reactivity worth, kinetics parameters, effect of device on pin power, in-core detector signal, radiation protection, depletion, etc.

The "production" MCNP model that is equivalent to the production WIMS-AECL/DRAGON/RFSP model for a CANDU full-core problem has also been developed. In order to obtain a reliable MCNP result, especially with regard to the power distributions, practical approaches are needed to remedy the above-mentioned issues in the MCNP modelling and simulation of the CANDU 3D full-core problem with limited computer resources [28].

11.7 Summary

The current deterministic codes (WIMS-AECL/DRAGON/RFSP) have matured and have been adopted for routine industrial applications, offering reliable tools for designing and operating CANDU reactors.

---

[2] The dominance ratio is a practical parameter relative to the ratio of the first two eigenvalues in critical nuclear systems. A value close to one is a challenging problem for neutron-transport simulations especially in the field of large thermal reactors.

In the area of lattice-cell and supercell calculations, WIMS-AECL and DRAGON are quite representative of the current state-of-the-art for production codes. In the area of 3D finite-core calculations, RFSP offers unique capabilities for the design and analysis of CANDU reactors.

As most of the Monte-Carlo codes (including MCNP) are still not sufficiently efficient to produce solutions of time-dependent problems for routine-production runs, the deterministic codes based on approximate methods will still be used for "standard" reactor-physics calculations in the immediate future.

12   References

1. J. R. Lamarsh and A.J. Baratta, "Introduction to Nuclear Engineering", Third Edition, Prentice Hall, Upper Saddle River, New Jersey, USA, 2001.
2. R.T. Jones, "Recommended Delayed Photo-Neutron Data for Use in CANDU Reactor Transient Analysis", AECL-CONF-348, Atomic Energy of Canada Limited, Canada. https://www.nrc.gov/docs/ML0236/ML023600317.pdf
3. B. Rouben, "The Essential CANDU – Chapter 21: CANDU In-Core Fuel-Management", Editor W. Garland, 2014. http://www.unene.ca/education/candu-textbook
4. "Science and Reactor Fundamentals Training Course", Volume 2, Rev. 1, Canadian Nuclear Safety Commission, Canada, 2003 January. https://canteach.candu.org/Content%20Library/20030101.pdf
5. B. Rouben, "Introduction to Reactor Physics", Training Material, Atomic Energy of Canada Limited, Canada, 2002 September. All content except tables and figures is available from https://canteach.candu.org/Content%20Library/20040501.pdf
6. D. Rozon, "Nuclear Reactor Kinetics", Polytechnic International Press, Montréal, Canada, 1998.
7. "CANDU 6 Technical Summary", CANDU 6 Program Team, Atomic Energy of Canada Limited, Canada, 2005 May.
8. C.M. Bailey, R.D. Fournier and F.A.R. Laratta, "Regional Overpower Protection in CANDU Power Reactors", Proceeding of the International Meetings on Thermal Nuclear Reactor Safety, Chicago, USA, August 29 - September 2, 1982.
9. J. Hu, D. Scherbakova, D. Kastanya, M. Ovanes, "ROP Design for Enhanced CANDU 6 Reactor", Int. Conf. Future of HWRs, Ottawa, Canada, October 2-5, 2011.
10. Werner Fieguth, "Reactor Control", Canadian Nuclear Society CANDU Safety Technology Course, 2018 March.
11. E.E. Lewis, "Nuclear Power Reactor Safety", John Wiley & Sons, USA, 1977.

12. P. Reuss, "Neutron Physics", EDP Science, France, 2008.
13. R. Roy, J. Koclas, W. Shen, D.A. Jenkins, D. Altiparmakov and B. Rouben, "Reactor Core Simulations in Canada", Proc. of the Int. Conf. on the Physics of Reactor (PHYSOR2004), Chicago, USA, April 25-29, 2004.
14. D. Altiparmakov, W. Shen, G. Marleau and B. Rouben, "Evolution of Computer Codes for CANDU Analysis", Proceedings of the PHYSOR2010– Advances in Reactor Physics to Power the Nuclear Renaissance, Pittsburgh, Pennsylvania, USA, May 9-14, 2010.
15. J.D. Irish and S.R. Douglas, "Validation of WIMS-IST", in Proceedings of the 23rd Annual Conference of the Canadian Nuclear Society, Toronto, Canada, 2002 June.
16. D. Altiparmakov, "New Capabilities of the Lattice Code WIMS-AECL," Proceedings of PHYSOR 2008, the International Conference on Reactor Physics, Nuclear Power: A Sustainable Resource", Interlaken, Switzerland, September 14-19, 2008.
17. G. Marleau, A. Hébert, and R. Roy, "A User Guide for DRAGON 3.06," Report IGE-174R7, École Polytechnique de Montréal, 2009.
18. A. Hébert, "Applied Reactor Physics", Polytechnic International Press, Third Edition, Montréal, 2020.
19. D.A. Jenkins, B. Rouben and W. Shen, "History of RFSP for CANDU Fuel Management and Safety Analysis", Proceedings of the 31st CNS Annual Conference, Montreal, Canada, May 24-27, 2010
20. W. Shen and P. Schwanke, "Evolution of RFSP 3.5 for CANDU Analysis", Proc. of the 33rd Canadian Nuclear Society (CNS) Annual Conference, Saskatoon, Canada, June 10-13, 2012.
21. B. Rouben, K. S. Brunner, and D. A. Jenkins, "Calculation of Three-Dimensional Flux Distributions in CANDU Reactors Using Lattice Properties Dependent on Several Local Parameters," Nuclear Science and Engineering, Vol. 98, pp 139-148, 1988.
22. B. Rouben and D. A. Jenkins, "A Review of the History-Based Local-Parameter Methodology for Simulating CANDU Reactor Cores,"

Proceedings of International Nuclear Congress (INC93), Toronto, Canada, October 3-6, 1993.

23. J.V. Donnelly, "Development of a Simple Cell Model for Performing History Based RFSP Simulations with WIMS AECL," Proceedings of International Conference on the Physics of Nuclear Science and Technology, Long Island, USA, October 5-8, 1998.

24. W. Shen, "Development of a Micro Depletion Model to Use WIMS Properties in History Based Local Parameter Calculations in RFSP," Proceedings of the Sixth International Conference on Simulation Methods in Nuclear Engineering, Montreal, Canada, October 13-15, 2004.

25. W. Shen, "Development of a Multicell Methodology to Account for Heterogeneous Core Effects in the Core-Analysis Diffusion Code," Proceedings of International Conference on the Advances in Nuclear Analysis and Simulation, PHYSOR-2006, Vancouver, Canada, September 10-14, 2006.

26. W. Shen and D. Altiparmakov, "Multicell Correction Method for Treatment of Heterogeneities in Full-Core Calculation of CANDU-Type Reactors", Nucl. Sci. Eng., Vol. 174, pp. 109-134, 2013.

27. X-5 Monte Carlo Team, "MCNP - A General Monte Carlo N Particle Transport Code, Version 5, Volume I: Overview and Theory", LA-UR-03-1987, 2006.

28. W. Shen and J. Hu, "Development of a Production MCNP CANDU 3D Full-Core Model with Practical Remedies to the Issues in Deriving Reliable Tally Results", Proceedings of PHYSOR 2016, Sun Valley, Idaho, USA, May 1-5, 2016.

29. L. Cao and H. Wu, "Deterministic Numerical Methods for Unstructured-Mesh Neutron Transport Calculation", Woodhead Publishing Series in Energy, Elsevier, 2020.

Bibliography

1. R.J.J. Stamm'ler and M.J. Abbate, "Methods of Steady State Reactor Physics in Nuclear Design", Academic Press, London, UK, 1983.
2. D. Rozon, "Nuclear Reactor Kinetics", Polytechnic International Press, Montréal, Canada, 1998.
3. P. Reuss, "Neutron Physics", EDP Science, France, 2008.
4. D.G. Cacuci, "Handbook of Nuclear Engineering", Springer Science & Business Media, Berlin, Germany, 2010.
5. A. Hébert, "Applied Reactor Physics", Third Edition, Polytechnic International Press, Montréal, 2020.
6. L. Cao and H. Wu, "Deterministic Numerical Methods for Unstructured-Mesh Neutron Transport Calculation", Woodhead Publishing Series in Energy, Elsevier, 2020.

www.ingramcontent.com/pod-product-compliance
Lightning Source LLC
Chambersburg PA
CBHW050501190326

41458CB00005B/1391